# THE GENOME ODYSSEY

# THE GENOME ODYSSEY

## MEDICAL MYSTERIES AND THE INCREDIBLE QUEST TO SOLVE THEM

**Euan Angus Ashley,** M.D., Ph.D.

CELADON
BOOKS
New York

 Printed in the United States of America. For information,
address Celadon Books, a Division of Macmillan Publishers,
120 Broadway, New York, NY 10271.

www.celadonbooks.com

Library of Congress Cataloging-in-Publication Data

Names: Ashley, Euan A., author.
Title: The genome odyssey: medical mysteries and the incredible quest to solve them / Euan Angus Ashley, M.D., Ph.D.
Description: First edition. | New York: Celadon Books, 2021. | Includes bibliographical references and index.
Identifiers: LCCN 2020042046 | ISBN 9781250234995 (hardcover) | ISBN 9781250234971 (ebook)
Subjects: LCSH: Genomics. | Medical genetics.
Classification: LCC QH447 .A84 2021 | DDC 572.8/6—dc23
LC record available at https://lccn.loc.gov/2020042046



First Edition: 2021

10 9 8 7 6 5 4 3 2 1

*To my mum and dad,*
*who gave me so much*
*and showed me what it really means*
*to care for other people*

# Contents

# Preface

I can't say for sure where my fascination with the genome began, but I'm pretty sure I know why it continues: unraveling the genome, our "code of life," is exhilarating. Day after day, we attempt to crack a code that is at once the essence of our humanity and also the destination for our pursuit of clues to help families afflicted with the most challenging genetic diseases. Life doesn't get more fulfilling than that.

None of this I could have imagined during a childhood growing up in the West of Scotland in the 1970s. That was around the time the first genome of a tiny organism was decoded. Since then, we've learned that your genome is a code linking you to every living organism on earth; a code incorporating the history of the human race; a code etched with your own family's history, dating back hundreds or even thousands of years. It is a code unique to you: no one alive and no one who has ever lived has the same one (not even your identical twin). Your code contains information about everything from your most likely height, weight, hair color, and eye color, to your predisposition, to some of many thousands of diseases. Your code can predict your future too: how you will live and how you may die.

All of this is inscribed by a molecule, deoxyribonucleic acid—DNA—

that has become so much a part of our lexicon that it has transfigured from molecule to metaphor. We say that a trait is "in the DNA" of a company or an institution despite the fact that neither of these is an organic being. We mean the trait is buried so deeply, bound so tightly, and wedded so irreversibly that it actually *is* the very fabric of its being. When we say of a fellow human that certain characteristics appear to be "in their DNA," we express the sense that these features are implicit, ingrained, and natural.

Paired in two strands, coiled into a right-handed double helix, DNA consists of just four simple building blocks arranged in patterns along its length. These nucleotides are the alphabet of life: A, T, G, and C—four letters forming the unique code that is locked in the vault of almost every cell in your body. Your genome is three billion letter pairs, six billion data points, two meters of molecule compacted into twenty-three pairs of chromosomes that, if laid end to end with the DNA from the thirty trillion cells in your body, could stretch to the moon and back thousands of times: part of the literal embodiment of what it means to be human.

The first human genome to have its letters spelled out (sequenced)—the Human Genome Project—was a ten-year, multicountry, multibillion-dollar effort to decode a mixture of DNA from ten people (actually, in the end, it was most of half a genome from more like two people, but more on that later). And while that first genome cost billions, it is, fewer than twenty years later, now possible to sequence a human genome for less than the cost of the cheap commuter bike I ride to work every day. This staggering reduction in cost has fueled a tsunami of scientific discovery and has given the medical profession an unparalleled opportunity to change lives for the better. Here is a tool for redefining disease, for solving medical mysteries, for providing hope to families who have suffered loss, and for protecting those who remain behind. This molecular microscope has allowed us to understand disease at a deeper level and to begin to personalize the practice of medicine.

The dramatic advances that you'll read about in this book—advances that have taken us from a theoretical understanding of DNA to the first human genome to a paradigm shift in medicine made possible by the sequencing of millions of genomes—came about because of revolutionary advances in our ability to read DNA and computer programs and human collaboration that have enabled us to understand it.

I think my younger self would have been surprised to hear I was dedicating my life to understanding the human genome even though, at some level, I was always going to be a doctor. Before the age of ten, I was the guy my school friends asked to wipe the blood off their grazed knees (my dad being the local doctor, I was the obvious choice, while my sister declared early as a veterinarian, so she got the dead birds). But I liked this role. I learned first aid around age twelve and became the resident expert among my peers on bodily functions. I bolstered this expertise by visiting my dad's medical office to inspect his tools and swivel on his doctor's chair. I was fascinated with his doctor's bag. It was black and box-shaped, came up to my knees, and appeared to me like a miniature hospital—a treasure trove of little drawers that opened to reveal needles, sutures, and intriguing metal implements, the utility of which I could barely imagine. I remember my dad once stitched a friend's forehead back together in our kitchen. That was extremely cool. Sometimes, he took me on his house calls. I remember one Christmas Day when he spent hours securing oxygen for one of his patients with lung disease to keep him out of the hospital. My mom was a midwife, and she would take me on her calls sometimes too. From both, I learned dedication, compassion, and that practicing medicine was not a job, nor even "just" a profession. It was a way of life, a state of being. I felt drawn to this on a visceral level: looking after people seemed like the thing I was put on this planet to do.

And yet, I was also definitively a technology geek. I remember winning $150 when I was about ten and being forced into Scottish frugality, "investing" the money in a savings account rather than buying a pair of spectacles with built-in windscreen wipers. My dad and I bonded over solving the Rubik's Cube when I was about twelve (I think my record was thirty seconds, but the rose tint is obscuring the details now). I do clearly remember that summer when I was supposed to be outside enjoying the Scottish "weather," I was instead to be found inside more focused on writing a computer program to calculate the payroll taxes for my dad's medical practice. Of course, they changed the tax code the next year, teaching me an important lesson about, well, governments and taxes. I wasn't destined for accountancy anyway (my brother is the one to trust with your money), but I did imagine great wealth for myself from the profits of a horse racing game I wrote for the Sinclair ZX Spectrum home computer when I was fourteen. You made your horse go faster by alternately pressing the rubber keys as fast as you could. It didn't make me rich, but my nerd friends thought it was awesome. Or so they said.

At school, that teenage geek was drawn not only to physics and mathematics but also to language and music. My biology teacher, who informed my parents on one occasion that I was a buffoon, unwittingly ignited in me, at age sixteen, a fire for genetics when he gave me a copy of Richard Dawkins's *The Selfish Gene.* (The same teacher also suggested, with only marginal humor, that perhaps my experimental apparatus had fallen on the floor before I drew it in my scientific notebook. In his defense, art was never my strong suit.) My new pastime became reading popular science books—Dawkins, Gould, Lewontin, Sacks, Diamond, Pinker, and, of course, Darwin. I carried this passion with me into medical school. One of our early physiology classes sealed my fate. In groups of four, we sat in a lab at low benches on uncomfortable wooden stools, waiting anxiously to receive a recently excised still-beating rabbit heart. We poked it, felt its squidginess, picked it up, marveled at its spontaneous beating, and hung it on a needle by the aorta, the blood vessel coming out of the top. Our job was to feed it, nurture it, and keep it alive as long as possible. I stared at this heart for hours. I was transfixed by its beauty—an elegant, exquisitely synchronized, biological machine. This wonder of evolution had a magical quality. It had rhythm and through its sounds produced music. I was hooked.

And so I found my medical calling. The specialty of cardiology had a vitality that was excitingly present: delivering electric shocks to patients whose hearts had stopped seemed to be an impactful way to spend one's day. And besides, many of my family suffered heart disease. Perhaps I could be useful to them one day.

Many years later, by virtue of a good dose of luck, I have found myself able to combine these passions as a practicing cardiologist and founding director of the Center for Inherited Cardiovascular Disease at Stanford University in California. The geek in me enjoys running a laboratory focused on using big computers and laboratory experiments to understand biology, and I have been fortunate, with talented colleagues, to cofound several biotechnology companies. As well as start-ups, I have been lucky enough to work with companies like Google, Apple, Amazon, and Intel, as well as the governments of several countries, advising them on the genome's place in medicine. I feel privileged to be alive at a time of technological transformation in genetics and health.

Today, you can sequence your genome for a few hundred dollars. Some years ago, I decided to devote myself to the cause of understanding the genome. It seemed to me that if we could parse its meaning, then perhaps we could start to decode our genetic future—to predict and prevent disease before it even began. I wondered if the quest might even inform, in some way, what it really means to be human.

This book tells of the first few years of adventures following that decision. Through its pages, I will take you on a journey into the science and medicine of the human genome. I will tell you stories of patients whose care has been transformed by knowledge of their genome. I will introduce you to scientific teams I have led and admired and trace a path from genome data to medical action.

The first few chapters ("The Early Genomes") introduce you to our team and our efforts to medically decode some of the first genomes to be sequenced. I describe wandering into a Stanford colleague's office one day in 2009 to find him perusing his own genome and how that revelatory moment led us to try to bring every then known human genetic observation to bear on his genome to better understand his risks of disease. I tell of his cousin's son, who died suddenly in his teens, and how we used genetic sequencing of his postmortem heart tissue to try to find answers for his family. I tell of the origins of the human reference sequence in Buffalo, New York, and of the high school student who brought the genome sequences of her entire family to school for a science project. I describe how we extended these efforts into our Stanford clinics and tell how we started a company to broaden the impact beyond those walls.

In the second part of the book ("Disease Detectives"), I try to convince you that genomic medicine is very much like detective work. We start with the scene of a "crime" and carefully observe clues and document evidence using the tools of traditional medicine as it has been practiced for thousands of years: observe, examine, document, analyze. To this, we add our newest tool, reading the genome. I tell stories of patients with undiagnosed disease who have spent years looking for answers and find them through the wonder of genome science. I describe a national network of disease detectives whose mission is to systematically end these diagnostic "odysseys."

In the third section of the book ("Affairs of the Heart"), I describe some of my most treasured cardiac patients. I talk about a budding Broadway star for whom a play of genetic chance left her facing the consequences of an enlarged heart and the prospect of a heart transplant. I describe how we tried to break genome speed records to diagnose and treat a newborn baby girl whose heart stopped five times on the first day of her young life. I relate how we found a young man with a big smile and multiple tumors growing inside his heart and how we traced the cause to a missing piece of his genome. I tell how we discovered a baby girl born with not just one but two different genomes. In this section, I also introduce you to the history of our understanding of heart disease and sudden death, the conditions that predispose to it, and the colorful characters whose work provided the insights we now use to treat it.

In the fourth and final part of the book ("Precisely Accurate Medicine"), I look ahead to the future. I discuss how examining superhumans—humans protected from disease by their genomes—can help make the rest of us just a little more "super." I describe several of the exciting new efforts to glean insight from sequencing millions of humans, including President Obama's Precision Medicine Initiative and the United Kingdom's Biobank. I talk about advances in curing and treating genetic disease, including genetic therapy, as well as methods to develop traditional drugs that are turbocharged by genomic insights.

No single book should attempt to cover all the ground, and so there are some important areas of medicine you will not read much about in this book. We do not dive deeply, for example, into cancer. As well as the fact that others have written eloquently on this subject, little of the amazing work in genetic testing of cancer patients and their tumors has involved sequencing the whole genome (most cancer tests involve deeply sequencing a subset of the genome). So the topic of cancer I have left to another day. I have also not included discussion about genetic tests during pregnancy. Our ability to pick up fragments of a baby's DNA in the maternal bloodstream has dramatically changed our approach to such testing, but these fragments are not routinely assembled into a whole genome, so I will hold on to those stories for now. Finally, there are many critical advances and compelling stories from my colleagues all around the world that I have not been able to tell in this book. I try to do justice to the stories of our own patients and our own contributions in trying to help them, but there is so much work from others without whose endeavor ours would not have been possible.

The heroes of this book are my patients and their families. Together, they share a burden of inheritance: the little kids who smile and laugh or cry through their clinic appointments; the courageous teens coping with new diagnoses and living out unanticipated realities; the parents who get up every day to tend to the needs of their differently abled child. They travel from one doctor to the next, living every day the worry of a future unknown. They are the reason I get up every morning. Our patients and their families never escape what the genome brought to their lives, and we can never tire in our pursuit of understanding their genomes more fully, finding ways to treat their diseases more effectively, and, when our efforts fail, hugging them more tightly and assuring them that we will not rest until we come together upon a brighter day.

# Part I

# THE EARLY GENOMES

# 1

# Patient Zero

*Today we are learning the language in which God created life.*

—UNITED STATES PRESIDENT BILL CLINTON

*We share 51% of our genes with yeast and 98% with chimpanzees*
*—it is not genetics that makes us human.*

—DR. TOM SHAKESPEARE, UNIVERSITY OF NEWCASTLE

It was obvious to Lynn Bellomi that something was very wrong. It was August of 2011, and Lynn, from the city of Arroyo Grande on the scenic California coast, had just given birth to a beautiful boy named Parker. At first, everything seemed normal, but after several weeks, she became suspicious. Parker was still having trouble with things most babies learn to do fairly quickly, like feeding and sleeping. He was only sleeping a few hours a night. He cried a lot. By March 2012, when he was six months old, he was missing milestones—not showing curiosity about objects around him and not rolling over, never mind sitting upright. He was referred to a developmental specialist, then to an eye doctor, then to a brain doctor, then to a geneticist. To make matters worse, by nine months old, Parker appeared to be having regular seizures. He underwent many scans and dozens of tests, including painful blood draws. No one could figure it out. "Constant appointments, constant driving," his mom recalled. "And it felt like we were doing all these things without a purpose." Months turned into years.

In 2016, when we first met Lynn and five-year-old Parker, they had been referred to our Center for Undiagnosed Diseases at Stanford, part of a national network of doctor detectives whose aim is to solve the most challenging cases in medicine. Much of the time, success comes from analyzing a family's genomes, those DNA instructions that are the recipe book for all our cells and systems. So, on June 28, 2016, we drew blood from Parker so we could derive DNA from his white blood cells and spell out every letter in his genome. We also did this for his mom and dad.

Three months later, on October 4, genetic counselors Chloe Reuter and Elli Brimble called Lynn to say we had found a genetic change in Parker that did not appear to be inherited from either her or Parker's dad. It was a brand-new genetic mutation that arose in Parker, and it appeared to disrupt a gene called *FOXG1*. Other patients with damaging variants in this same gene had health problems that were remarkably similar to Parker's. This had to be the answer. For the first time since she had detected a problem with Parker's development, five years before, Lynn understood the size and shape of the enemy. She instantly gained a support group of families suffering with *FOXG1* syndrome around the world (650 parents on the *FOXG1* Facebook group, at last count). More than that, finally understanding the cause of Parker's disease allowed us to refer him to a movement disorder specialist who immediately changed his medications in a way that dramatically reduced his symptoms. "He still has some seizures, but much less frequently now," his mom recently told me. "He still has to go regularly to the doctor, but otherwise, he's a very happy guy."

Parker and his parents can now look forward to a world of new possibilities: joining with doctors, scientists, and hundreds of families from around the world to attack this disease from every angle, share experiences, disseminate insights, and hopefully, one day, find a cure. That future would have looked very different if it weren't for advances in our understanding of the genome—discoveries made, over the last few decades, by scientists whose work has had a profound impact on the way we detect and treat human disease. To explore those breakthroughs, let's start by going back to 2009.

It was a pretty ordinary day. I had completed my morning meetings, and instead of lunch, I was heading to the office of a future friend, a Stanford physics professor and bioengineer named Stephen Quake. Steve was well known for his pioneering work in the field of microfluidics. He invented tiny biological circuit boards with switches, kind of like railroad "points," to direct cells or molecules to a specific micro-destination for analysis. Steve and I were meeting to plan an afternoon symposium for the genetics faculty at Stanford. His office is in a building at Stanford named for James H. Clark, an electrical engineer and founder of Silicon Valley companies like Silicon Graphics and Netscape. Designed by the famous British architect Norman Foster, the Clark building is shaped like a kidney with swooping red lines and lots of glass. At night, it is brightly illuminated and looks for all the world like an alien spacecraft that landed in the middle of campus. In a way, it kinda did. The building's purpose was to gestate a new specialty—bioengineering—the love child of a dalliance between biology and engineering. Situated on campus right in between the schools of medicine and engineering, against a California landscape of blue sky, sunshine, and palm trees, it is a stone's throw from the Stanford hospitals. Through its windows, as you walk by, you see brightly lit rows of worktops harboring the trade tools of engineering right next to the wet benches of molecular biology—robots interbreeding with pipettes. And during the day, after navigating a curious and gratuitously complex room numbering scheme, if you're lucky, you find Steve's office on the third floor.

Steve is the archetypal physics professor—Stanford and Oxford trained, a brilliant iconoclast. The breadth and diversity of his intellect emanates from a brain enveloped by tufts of professorial hair that, in a former era, would be grown wildly to match his imagination. In fact, Steve's office is set up very much as I imagine his brain to be—mountains of chaotically "organized" scientific papers are piled up on every side and in every corner. He sits hunched in the middle, pecking at a keyboard, the creative energy source powering everything around. Amid a campus of overachievers, Steve stands out. I had gone that day to talk about a seminar we were running to bring together human geneticists across campus. But we never really got to that.

"Come look at this," he said. I found a place to sit amid the piles of journals, and he beckoned me over to look at the screen. It was not obvious at first what he was pointing at. There was a web browser open and a table filled the screen with the word *Trait-o-matic* at the top. It was one of those bare-bones spreadsheets with no formatting that was found on early websites—not pretty, but it was not the aesthetic that drew me in; it was the content. There were lots of columns of data. Gene names, gene symbols, As, Ts, Gs, and Cs, the building blocks of the genome.

"So what is that?" I asked.

His answer was to mark a pivotal moment for both of us. Delivered matter-of-factly and with more than a hint of his trademark understatement, it was as low-key as it was utterly revolutionary:

"It's my genome."

To put this into context, this was early 2009, and you could count on the fingers of one hand the number of people in the entire world whose genomes had been sequenced. Each one had been marked by an order of magnitude or more in reduction of cost. The Human Genome Project had been funded for $3 billion by the Department of Energy and the National Institutes of Health (NIH). And while each subsequent effort saw steep declines in cost, the price tags were still staggering. Craig Venter, the renegade entrepreneur who had taken on the public genome project in a race to be the first to sequence a human genome, sequenced his own genome at a cost of around $100 million. An anonymous Han Chinese man had been sequenced in 2008 for around $2 million. And James Watson, who shared the Nobel Prize for work with Francis Crick and Maurice Wilkins and who, together with Rosalind Franklin, elucidated the structure of DNA, had his genome sequenced by a group at Baylor College of Medicine in early 2008 for the comparatively modest sum of only $1 million. Each of these projects involved hundreds of scientists and thousands of hours of time, as well as no small amount of blood, sweat, and tears. Then in 2009, Steve sequenced his own genome, in his lab, using a technology he invented himself, with postdoctoral scholar Norma Neff and Ph.D. student Dmitry Pushkarev, for just $40,000. In one week.

I was familiar with sequencing both in my research lab and in my clinic. We would send blood from our patients for DNA sequencing as a medical genetic test to try to find the cause of their inherited heart disease. Those tests would spell out the ATGC letters of the five to ten genes we knew could cause their heart condition to try to find the (usually) one-letter change that was the culprit. At that time, the cost for sequencing these five to ten genes was around $5,000 and the results took two to four months to come back. The test would provide an answer only about a third of the time, given that we were still in the early days of matching genes to diseases. So that was my context. To imagine we might have access to a whole genome—not five, not five hundred, not five thousand, but all *twenty thousand* genes as well as the other 98 percent of the genome that falls in between the genes . . . well, that was simply mind-blowing.

A few of us had, at that time, started to wonder, with the steep decline in the cost of genome sequencing, if one day patients might walk into our offices, figuratively or literally, "clutching their genomes." In Silicon Valley, we like to compare everything to computers, but the parallel between the rapid decline in the cost of sequencing and the rapid decline in the cost of computing power was an enticing metaphor for many even beyond California's Bay Area. It became common among scientists to compare the drop in the cost of sequencing to Moore's law. Gordon Moore was a native of the Bay Area—a physicist who, along with Robert "Bob" Noyce, made fundamental contributions to the development of the integrated circuit, not least by starting Intel, one of the foundational semiconductor companies of Silicon Valley. In an article in 1965, referring to the rapid pace of technological advance, Gordon Moore observed that the rate at which components could be added to integrated circuits was doubling approximately every year, meaning the price of computing power was halving in this same period. He later decided two years was more realistic, but regardless, this "law" became synonymous with rapid technological advancement. It became common to show that the price of sequencing was dropping at a similarly spectacular rate, at least up until 2008, when the precipitous rate of decline in the cost of sequencing left Moore's law in the dirt. The National Human Genome Research Institute famously illustrated this by releasing a graph with a steep cliff-like falloff. I liked this graph and, like many genome researchers, would show it in my presentations. However, I soon found a more concrete, visceral way to put the price drop in perspective. My commute, at the time, took me past the Ferrari-Maserati dealership near Atherton—billionaire territory in the heart of Silicon Valley. I would often cast a sideways glance at those cars as I waited in traffic. One day, I was sitting at the stoplight doing random math in my head, as one does, and realized that if the Ferrari in the window had dropped in price as much as human sequencing had dropped in price in the eight years since the Human Genome Project's draft sequence was released, instead of $350,000 it would cost less than forty cents. A forty-cent Ferrari! A millionfold reduction in price. That seemed unprecedented. So I added that image to my slideshow. Sometimes, people tell me it's all they remember.

Admittedly, in 2009, with the cost of Steve's genome at $40,000, the notion that patients would start bringing their genomes to clinic still seemed like a preposterously futuristic scenario, about as likely as my owning one of those Ferraris. But futurism is a potent driver of creative ideation. Shouldn't we start preparing for that day? Yes, there would be computational challenges and huge gaps in knowledge. But if we could meaningfully decode the genome, not just sequence it; if we could not just read the book but actually *understand* it; if we could turn data into knowledge, then put that to work for patients? Whoa.

So there I was in Steve's office, and he was asking me about various genes and pointing on the screen to the places where his own DNA letter was different from the one in the reference sequence (we will discuss the reference sequence and where it came from in chapter 6). "Do you see anything you recognize?" he asked. I scanned the names and noticed a gene I knew really quite well: cardiac myosin-binding protein C. This gene encodes for a protein that is an important part of the molecular motor of the heart. Its true function eluded scientists for years, but we now know that variants in this gene are the most common cause of the inherited heart disease hypertrophic cardiomyopathy—a disease associated with heart failure and sudden death. And here was Steve pointing to a variant in his genome in that very gene. There was a chance such a variant could be life-threatening. So naturally, being a cardiologist, I started asking him about his medical history. Do you have any medical conditions? Any symptoms? Chest pain? Shortness of breath? Palpitations? Instead of the scientist who walked into a colleague's office, I was now the physician talking to a patient: a very different kind of investigator, probing a very personal kind of truth. To my relief, Steve had not experienced any such symptoms and had no known medical condition.

So I moved to his family history. *Family history* means such different things to different doctors. To some, necessity requires it be a checkbox question: "Nothing in the family?" Move on. But to a geneticist or diagnostician of rare disease, the family history is a treasure trove, a box of clues, to be scoured through, picked apart, examined, and deconstructed. This brand of diagnostician treats the family history like Sherlock Holmes treats the crime scene: examining it in minute detail, from every angle, actively interrogating, then reflecting. Yet few of us really know our family medical history well. Try it right now for yourself. Make a list of diseases that run in your family, and then try to match the names of the relatives who suffered from each disease along with how old they were when they were first diagnosed. Not so easy. I asked Steve the question, and like most patients, he answered quickly, "No, no family history of disease." And then, as if accessing some dusty file at the far end of the cabinet, "But wait, my dad has some heart thing, a rhythm problem . . . ventricular . . ."

"Tachycardia?" I offered, not expecting that to be right but, rather, reacting instinctively with the worst possible scenario (it's a doctor thing). Ventricular tachycardia is an abnormal heart rhythm that can occur, for example, in hypertrophic cardiomyopathy.

"Yeah, that sounds right."

Well, now, my curiosity was tinged with concern. In ventricular tachycardia, the normal coordinated activity of the top and bottom chambers of the heart is replaced by a rapid, dangerously uncoordinated rhythm that can be ineffective at pumping blood. Low blood flow to the brain means no consciousness and a rapid end to life. It is a rhythm that strikes fear into the heart of most doctors because it is almost always a medical emergency. Doctors *run* when they are called to patients in ventricular tachycardia. The name itself seems to lay down a staccato rhythm that evokes the broad, unruly electrical signal seen on the hospital monitors. It is a rhythm that screams, "Act now!" Fast, feared, and sometimes fatal.

So to recap, I had wandered into this meeting about organizing a genetics seminar, and here was my new friend Steve, the world-famous scientist, telling me that his dad possibly had reasons to suffer from ventricular tachycardia, a condition associated with sudden death. And here was I, a cardiologist specializing in inherited cardiac diseases causing sudden death, staring at his genome at a particular variant in a particular gene known to be associated with hypertrophic cardiomyopathy, an inherited cause of sudden death. "So has anyone in your family ever died suddenly?" I asked. This question is arguably our most powerful tool. Such questions and their follow-up are, to a physician, as surgical tools are to a surgeon. Each surgeon has their favorite tools, some even personally crafted. They feel right in the hand. The balance is just so. The surgeon knows how her favorite tool responds, how it cuts. She understands innately the tissue response. When wielded in the right way, questions like this are the diagnostician's scalpel.

"Well, actually . . . my cousin's son recently died suddenly, and no one knew the cause."

Boom.

There it was: a family history of sudden unexplained death. The reddest of red flags, unfurled in front of me, waving in my face. Trying to sound casual, as I less-than-casually performed the mental math to calculate Steve's likelihood of sharing a genetic condition with his cousin's son, I said, "Oh, really, what sort of age was he?"

"Oh, he was only nineteen, a black belt in karate, I think, and never had a sick day in his life."

He had my attention now. The most common causes of sudden death in the young are inherited cardiac diseases like cardiomyopathy, like hypertrophic cardiomyopathy. And as I invited Steve to come to my clinic so we could check his heart, it was in that moment he became not just my colleague and my friend but my patient. And about a nanosecond after that, as my mind raced to work out how to put this together—how fast and what favors I would need to pull to screen his heart without delay—I realized that he was also about to be the first patient in the world to walk into a doctor's office for a checkup with his genome.

His complete genome.

And the doctor was me.

I headed back to my office, my mind racing with the possibilities and the impossibilities. How would one go about analyzing a genome, anyway? At the time, the idea of interpreting a whole human genome seemed as premature as it was preposterous. The handful of genomes released publicly at that point had undergone analyses that were mostly represented in statistics—this many single-letter variants were found, for example. The team at Baylor had gone further and looked at variants in medically relevant genes in Jim Watson's genome. But to think about scaling a medical approach to a whole genome, including every variant in every gene, was just not something anyone we knew had a viable solution for yet.

So I found one of my cardiology trainees, now a long-term collaborator and friend, an extraordinarily talented clinician-scientist named Matthew Wheeler. Matt is originally from upstate New York and trained in Chicago before coming to Stanford. He is tall, broad enough to power a rowing boat at speed, and nimble enough to look a whole lot better than I do skiing down a mountain. Our meeting was in fact orchestrated by our wives at a crew party for their rowing club, and we found shared passions in cardiology, genetics, sports, and inherited cardiovascular disease. On that day, we spoke about an ambition to build a Center for Inherited Cardiovascular Disease. This day, five years later, when we met in my office (which later became his office), I told him about Steve, his genome, his family history, and the idea that had formed in my head since walking back from my meeting: the idea, well, of clinically analyzing a whole human genome, every position, every gene, every variant. His response was typically understated, delivered deadpan, almost sotto voce, and foretold of the adventure upon which we were about to embark:

"Glad to see ambition hasn't left the building."

. . .

The human genome lives inside almost every cell in the body. I say "almost" every cell because certain cells, like red blood cells, for example, lose their nucleus as they mature—all the more room to pack in oxygen. The genome is housed in the cell's "inner vault," the nucleus, although some genes also live in the "power units" of the cell, the mitochondria. As mentioned earlier, the genome is made up of extremely long molecules of DNA. The individual chains of DNA are long strings of molecules called *nucleotides,* special sugars with one of four bases attached. The bases are adenine, thymine, guanine, and cytosine. The initial letter of each of these bases—ATGC—makes up the genetic code, six billion letters long. The DNA molecules that make up the genome are so long that if you were to stretch out the DNA from just one cell, it would be two meters long. That DNA needs to be compacted so it can fit into the nucleus. To achieve this, it is wrapped around proteins called *histones* and packaged into a compact structure called *chromatin* that makes up individual chromosomes. Each normal human genome has twenty-three pairs of these chromosomes: twenty-two regular ones and one pair of sex chromosomes, combinations of X and Y (females have two X chromosomes, males have one X and one Y). Some diseases emerge from duplications of whole chromosomes; for example, trisomy 21, a condition also known as Down syndrome, occurs if you have three copies of chromosome twenty-one. So to recap, the genome is a recipe book contained inside almost every cell in your body. It's six billion letters, all of them A, T, G, or C, and compacted into chromosomes of which most people have twenty-three pairs.

This recipe book contains ingredients and instructions for what to do with them. The ingredients are genes. These vary enormously in size: the smallest one is only eight letters, while the longest is 2,473,559 letters. Most genes represent the instructions for building a protein. To get there, the DNA is transcribed into a related molecule called *ribonucleic acid* (RNA) that carries the code as a message out of the nucleus for it to be translated, in groups of three letters at a time, into amino acids, the building blocks of proteins that do the work of the cell. Proteins can be structural, to hold the cell together, or motor proteins, to move themselves or other things around, or enzymes that can convert one molecule to another. Yet the twenty thousand or so genes that account for all these proteins make up only about 2 percent of the genome. What about the other 98 percent? It seems almost unfathomable now that, for many years, this part of the genome was referred to as "junk DNA," reflecting the fact that nobody really knew what it was for. Our naïveté in assuming nature had no use for the vast majority of our genome embarrasses us further every year, as we learn more of the secrets of this "dark" genome. It turns out that this noncoding part of the genome is vital in determining whether genes are turned on or off. Also, about half of our genes have associated pseudogenes in this part of the genome—copies of the gene that are no longer functional (or so we used to think—now we know that pseudogenes can also regulate the function of other genes, especially their partner gene). Some of it sure *looks* a lot like junk; half the genome is made up of repeating sections of DNA that we still don't really understand. Finally, and perhaps craziest of all, almost 10 percent of our human genome is actually derived from viruses that embedded themselves long ago in our genome. Remember that the next time you have a cold.

Deciphering something as complex as the genome is something that seemed impossible for many years. In the 1970s, two approaches to reading DNA were proposed, but the one invented by Frederick Sanger came to dominate. Sanger was a British biochemist who, despite being one of only four people ever to win the Nobel Prize twice and mentoring two Ph.D. students who themselves won Nobel Prizes, used to describe himself as "just a chap who messed around in a lab." Sanger's approach, which dominated sequencing for decades and still plays a major role today, takes advantage of a molecular copying machine present in all our cells called *DNA polymerase.*

To understand Sanger sequencing, we're going to get a bit technical for a minute. Think about starting with four tubes labeled A, T, G, or C. In each, we place our DNA copier, the DNA molecules we want it to copy, and the building blocks for making DNA (As, Ts, Gs, and Cs). Now, to each tube we add a special version of just one building block, the one corresponding to the label on the tube. The building block has a radioactive flag attached. When incorporated, this flag prevents the copier from lengthening that particular DNA molecule any further. Also, importantly, we add just a little of it compared to the regular building blocks. Now imagine as the copier in each tube starts, it grabs the building blocks it needs for the DNA molecule it is copying randomly from the mixture. Of course, it has a higher chance of incorporating regular building blocks than the special building blocks, because there are so many more regular ones. Eventually, however, by chance it grabs one with a flag. At that moment, the DNA copier is stopped in its tracks, and that molecule is flagged as radioactive. The copier moves on to make new copies elsewhere in the tube, and the cycle repeats. Eventually, there are four tubes each containing DNA copies of variable lengths. The A tube contains copies tagged with an A. The T tube contains DNA copies tagged with a T, and so on. To read the sequence, the DNA from each tube is taken out, and the molecules are spread out according to their lengths down a slab of gel using electrical charge. The radioactive elements can then be detected by exposing the gel to photographic film. The result is four tall, thin photographs, each of which appears like a ladder with lots of rungs missing. However, here's where the magic happens. If you line up the four photographs next to each other, you see that each rung is represented in only one of the photos. The ladder in which the rung appears corresponds to the letter for that position: A, T, G, or C.

If you didn't catch all of that, just bear with me. This laborious process was accelerated and commercialized with three major advances: 1) radioactivity was replaced with light-emitting molecules, 2) everything was run in one tube, and 3) molecules were separated much faster and more efficiently based on their electrical charge. This technology, reading DNA copies each about five hundred letters long, was developed by the company Applied Biosystems and became the workhorse sequencing approach for the Human Genome Project.

The second genome, completed using this same technology, was finished around the same time as the Human Genome Project and belonged to Craig Venter, a scientist who had formed a company to sequence and attempt to patent human genes. He created a firestorm by challenging the public program to a race to the finish (in the end it was declared a draw). Venter's genome cost around $100 million to sequence (representing a staggering drop in price for our Ferrari from its original $350,000 to a mere $12,000).

Many such breakthroughs in biology come to parallel, if not the content, then certainly the language of science fiction. And that may, or may not, be the reason that so-called next-generation sequencing was born. *Star Trek*'s Jean-Luc Picard would have been proud. And of course, since *next* is a relative—not an absolute—term, it was perhaps inevitable that almost everything since Sanger sequencing has been referred to as *next generation* at one point. It is truly the gift that keeps on giving. And that gift is confusion. But the thing that all next-generation technologies have in common is their ability to expand the process of sequencing. Instead of focusing on the part of the genome you want to sequence, making many copies of just that part, then Sanger sequencing those, with next-generation sequencing, you take the whole genome, chop it up into small pieces around one hundred letters, then sequence all the pieces at the same time. This allows, effectively, a massive turbocharge to sequencing.

Such a technological advance took some time to finesse. It was to be seven years before another individual's genome was published. In 2007, the Nobel Prize winner James Watson's genome was sequenced using a technology from a company called 454 (founded by serial entrepreneur Jonathan Rothberg) by a team at Baylor College of Medicine led by

Australian geneticist Richard Gibbs. Roche bought the mysteriously named 454 technology in 2007 because of its ability to sequence very long pieces of DNA (initially four-hundred- to five-hundred-letter fragments; it was later updated to read fragments up to one thousand letters). According to Baylor's analysis, Watson's genome revealed a predisposition to cancer. He also famously redacted from the public disclosure of his genome the status of a gene variant predisposing him to Alzheimer's disease. His genome took two months to complete and cost $1 million. That Ferrari was just discounted to $116.

Various groups around the world published three more genomes (these were anonymous) in quick succession at the end of 2008 and beginning of 2009. All were sequenced using technology from a company called Illumina, the dominant force in sequencing for most of the last ten years. Importantly, these genomes started to represent more of the world's diversity: one individual was Han Chinese, one was Korean, and the other was West African. The last publication included some medically oriented annotation of the genome and even used an early version of the Trait-o-matic software that I first saw in Steve's office. Each took about six to eight weeks to complete and cost a few hundred thousand dollars: a trio of $50 Ferraris.

Steve's genome stood out for several reasons. For starters, he invented the technology used to sequence the genome and founded a company, Helicos, to market the instrument he invented—the cutely named HeliScope. The Helicos approach differed from the Sanger and Illumina techniques, because it sequenced single molecules of DNA. Fluorescently labeled DNA bases were flooded into a device called a flow cell where the short target stretches of DNA were anchored. As each base was incorporated into a new strand of DNA by DNA polymerase, the copier, a very sensitive camera, would take a picture, kind of like taking a photo of a tiny light bulb. Then, a wash step would chop off that light bulb, and another would be flowed in. Then another photo would be taken, and the cycle would repeat. But of course, each photo wasn't just one light bulb. The camera could read a billion bulbs at once, meaning that enough data to cover a whole human genome could be generated *in one week,* at a cost of $40,000. Today, sir, your Ferrari will be assembled in just under one hour and discounted to $6.

As you might imagine, all these next-generation approaches output millions of short genome "words" that correspond to the small fragments of DNA that are fed into the sequencer. The words do not come out in any particular order, so to be understood, they need to be organized—put together kind of like a jigsaw. That is usually done via a computer program that scans the human reference sequence (the sequence created by the Human Genome Project) and locates the correct position for each new word. Such programs are standard now, but at the time, the software had to be written from scratch. That job fell to Dmitry Pushkarev, from Steve's lab, a tall, lean Russian graduate student with enviable stamina both in late-night coding and daytime adventure pursuits. Dmitry built some of the first programs to stitch together a genome and find the places where it varied from the human reference sequence. And it was in that data and those algorithms that our work began.

# 2

# Team of Teams

*It is amazing what you can accomplish if you do not care who gets the credit.*

—HARRY S. TRUMAN

*None of us is as smart as all of us.*

—KEN BLANCHARD

To tackle a challenge as daunting as analyzing a whole genome was clearly going to require a large and diverse team. To assemble the key players, we first had to think about what we needed that team to do. There are broadly three categories around which we think about human genetic variation in relation to health and disease, and we needed champions for each area. First, there are large disruptions to the genome. These kinds of disruptions are rare and cause disease with little contribution either from other genetic variants or from the environment. At the extreme end, a whole chromosome can be duplicated, but more commonly, diseases are "single gene" disorders like cystic fibrosis, where rare variants in just one gene are largely responsible for the disease presentation. Variants like these disrupt the normal functioning of one or both copies of a gene sufficient to cause disease in most people, eventually. Variation between people with the same variant is caused by small contributions from other genes or the environment. Another category of genetic variation includes variants that are common in the population;

some are perhaps even found in the majority of people. These variants, as you might guess, almost never cause disease on their own because, individually, their effects are very small. But if hundreds or thousands or millions of them act together, that can add up to significant risk (examples include high blood pressure or heart attack risk). A final category, really a subset of each of the first two, comprises genetic variants that change the effects of medical drugs, a specialized area called *pharmacogenomics.* So these were the categories in which we needed champions. But before we could even get to thinking about who our champions might be, we had to consider the ethics of what we were considering. And fortunately, I knew just the person to tap for that.

In addition to his legendary erudition, Henry "Hank" Greely is known for his big personality, his ready candor, his colorful sweaters, and his unilateral rejection of PowerPoint slides. He is a specialist in the ethical, legal, and social implications of new biomedical technologies—especially genetics—and is an imposing presence in any room. I knew Hank by reputation, and we had first discussed the idea of what it would take to medically analyze a patient's whole genome some time earlier, following a meeting where he was kind enough to offer an ethics consult on a thorny patient problem. We talked more during a small group session at a Stanford genetics symposium a few weeks later. Actually, it was the same symposium I'd been planning to discuss with Steve Quake that day in his office when he first showed me his genome. In one part of this meeting, we were seated in small groups in a big room, imagining what it would take to really analyze a whole genome. I remember, vividly, Hank describing his vision of a stoplight approach to the predicted health effects of genetic variants: green—ok; yellow—take care; red—watch out. It was an intriguing idea. He suggested again that perhaps someone should try to do this for every variant in a whole human genome. What a crazy idea. In fact, that was exactly the plan that was now taking shape.

As it happened, at the very same meeting, there was someone else

I was looking to co-opt. Atul Butte is often described as a rock star of his field. This is due less to his excellent singing voice and intuitive grasp of the twelve-bar blues and more due to his ability to deliver high-energy, high-impact talks to packed arenas more befitting U2 than your average closed-door scientific meeting. Atul came to Stanford from Harvard, where he had trained as a physician in pediatrics and completed his Ph.D. under the famous Harvard scientist Zak Kohane. He ran a large lab with many high-energy, creative young data scientists. Of interest for our purposes, he had also been assembling a database of common genetic variants studied by other scientists for their relationship to disease. Much more important than simply gathering the data for this database, however, was *curating* the database. The distinction was key since, at the time, there was no automated way to extract data from scientific papers. His team read every single original paper, found the evidence linking the variant and the disease, then notated it in a standard format along with other details of the study like the number, gender, and ethnicity of the study participants. Cleaning the data in this way makes it much more computer-readable, or "computable"—something that would allow us to rapidly analyze a single whole genome and determine which variants were present. Atul, true to his exuberant personality, embraced the ambition of the idea and immediately signed up his whole team, including three very talented young data scientists.

We also needed to get a handle on genetic variants relevant for medication prescribing, such as those that make a patient particularly sensitive to certain drugs. In many ways, pharmacogenomics is what people think of first when they think about personalized medicine. And yet, when we go to our doctors, we get the same drug for high blood pressure that was prescribed for the last patient and the one next in line. We don't have the same diet, we don't have the same physiology, and we certainly don't have the same genome. So why would we automatically get the same drug? Instead of giving every patient the same medication at the same dose, what if your doctor could actually peer into your genome and choose the medication and dose that was right for you? For many years, electronic medical record systems have been able to look up your existing medications when your doctor prescribes a new one to check for possibly dangerous interactions between these and the new drug. We started to think that these systems could be co-opted to also look up your genome. If we were going to properly explore the entirety of a patient's genome from a medical perspective, we certainly needed an approach to drug therapy. What we needed was a database of pharmacogenomic variants. Well, as luck would have it, there was Russ Altman.

Russ is the living embodiment of the word *enigmatic*. He is one of those people whose name you can mention to almost anyone at Stanford and you will immediately prompt a smile. Some people walk to meetings; Russ bounds. Some people sit still in meetings; Russ fidgets. Brimming with creative energy, he is almost always first to leap to the whiteboard. Edward Feigenbaum, a pioneer of artificial intelligence, was my neighbor on the Stanford campus for a while. He remarked once to one of my other neighbors that he thought Russ was one of the smartest Ph.D. students he'd ever met. When we started work on Steve's genome, Russ had risen to become cochair of the bioengineering department, in partnership with Steve. In addition to that day job, he led the bioinformatics Ph.D. program and the Pharmacogenomics Knowledgebase project, known as PharmGKB. This was, and remains to this day, the bible for pharmacogenomics. Every important genetic variant known to be linked to medication response is listed and explained there. The team, run by Russ and his long-term collaborator Professor Teri Klein, use a combination of human and artificial intelligence to derive relationships between drugs and genetic variants from the medical literature. The result is an exquisitely curated resource, a computer-readable form of all the information relating genetic variants to medical drugs in the world. In other words, just exactly what we needed.

So the team was formed. A team of teams, in fact, covering the major areas of genetic health information we hoped to mine in Steve's genome. Steve's team would provide the list of variants, my team would look at rare variants and rare diseases, Atul's team would look for common variants for common disease, Russ's team would look at variants affecting drug responses, and Hank's team would keep us on the right ethical track. I reflected on how such an incredible collection of the brilliant, the pugnacious, the opinionated, the collaborative all had, without a moment's hesitation, committed to one of the craziest ideas they'd ever heard. Analyzing a human genome? Ingesting and understanding the totality of knowledge on human disease genetics, then applying it to the genome of one individual? In answer to such a ridiculous question, their unanimous answer was, "Tell me where to be and who to bring."

This team of teams was about to become friends.

Before digging around in Steve's genome to unearth risks for disease, including whether he carried any variants that might lead to a risk of sudden death, our first order of business was to engage Steve as "the patient" in a process of informed consent. We determined this needed to include a component of genetic counseling: While Steve undoubtedly understood genetics well, did he truly understand the *medical* implications of what we might find? For that matter, did we?

Genetic counseling is a profession with a relatively short but vital history. Genetic tests, unlike other medical tests, often have implications beyond the individual. Genetic tests need to be understood in light of family history and always carry relevance for other family members. They can be complicated to understand, rarely produce black-and-white results, and are used in medical situations where psychological support for the patient and family can be critical. This is where genetic counselors come in. Genetic counselors are a combination of counselor, educator, detective, technical validator, technical translator, psychologist, and therapist (and there is a global shortage, if you're looking for a job). At that point in 2009, genetic counselors had been concentrated mostly in prenatal care, rare disease medical genetics clinics, and a small number of subspecialties, including cancer and cardiology. In each of those settings, with few exceptions, genetic counseling prepared a patient for focused genetic testing—focused, that is, entirely on the problem at hand. Have you inherited your family's risk of breast cancer? Will your baby be at risk of a genetic disease due to a genetic contribution from each parent? After all, if you can only test a small handful of genes due to cost or there is only one gene known to cause the specific disease for which you are testing, the value of the test is clear, as is the range of possible results. But what if instead of a narrowly focused test, you had a test for "everything"? You suddenly have to face the possibility that if you test for "everything," you might find a lot of things you weren't looking for. What if we cleared Steve of known variants for sudden cardiac death but found an elevated risk for cancer? What if we discovered he was at risk of early-onset Alzheimer's disease (the very thing James Watson did not want others to know)? What if we found he was at risk for a condition with no known treatment? What if we found out that he had a variant that carried minimal risk for him but potential risk for his daughter if she inherited the variant? There was no playbook for any of these scenarios. We were in entirely new territory. In fact, we didn't even have a vocabulary for this. In a famous commentary in the *Journal of the American Medical Association* in 2006, Zak Kohane of Harvard Medical School along with Russ Altman coined the term *incidentalome,* a play on the words *incidental* and *genome.* They were referring to the incidental variants found when testing at genome scale, and they considered them a threat to advancing the field, thinking the uncertainty around them would scare people off from getting genome-scale testing. Incidental findings were, of course, not new to medicine. A heart murmur picked up while listening for signs of pneumonia in the lungs is an incidental finding that has been popping up for hundreds of years. A notch on a rib eventually diagnosing cancer might be seen incidentally on a chest x-ray looking for signs of chronic lung disease or asthma. Should genetic incidental findings be reported to the patient if there is uncertainty over what they mean (for very many variants, we don't have enough information on them to know if they are dangerous or not)? Could a patient not familiar with genetics truly give informed consent? For that matter, how could we provide advice to a patient regarding what we *might* find, when we really don't know what we might find?

Fortunately, in our ethics team, we had an experienced genetic counselor, a professor who also ran the master's degree program in genetic counseling at Stanford, Kelly Ormond. We approached Kelly with this unusual request. In some ways, our patient turned out to be easier than future patients would be. Steve, an inventor of genomic technology, wanted to know everything. Untreatable? Yes. Uncertain? Bring it on. However, we were thinking through the subtleties of informed consent for genome sequencing in the context of real patients for the first time, and many renowned genomic scientists like Steve actually have no exposure to the range of medical conditions a geneticist might see in clinic. We were planning an initial meeting with him to go over our approach and show some early results. Kelly immediately homed in on the rather unusual nature of the situation where Steve was both researcher and patient. She dived in, working with Hank to consider carefully all the ramifications of this dual role.

Meanwhile, it was time for Steve to visit the cardiology clinic. He did, after all, have a very significant family history of cardiovascular disease, including ventricular tachycardia in his father and sudden death in his cousin's son. In comparison with what lay ahead in looking at his genome, this part seemed easy. A few years before, I had started the Center for Inherited Cardiovascular Disease at Stanford with my nursing colleague, someone who became a work partner and close friend—an energetic, empathetic, and eternally cheerful morning person—Heidi Salisbury. A fervent patient advocate, indefatigable nurse leader, skillful soccer player and coach, Heidi is one of those nurses you want on your patient's side. We have looked after patients and families with all forms of inherited cardiovascular disease together for more than a decade, and it was to meet Heidi that Steve wandered nonchalantly into our clinic on a warm fall day.

While genome technology was undergoing a revolution, the more traditional tools of medicine had been changing little since René Laennec first rolled up a quire of paper into a cone to avoid putting his ear on the chest of a young woman in 1816. The stethoscope remains a straightforward device to amplify internal sounds from a patient's body to the ear of the doctor. Fortunately, I heard nothing abnormal when I placed my stethoscope on Steve's chest, just a *lub* and a *dub,* regular, rhythmic. Next was the electrocardiogram.

The electrocardiogram is a slightly more modern invention—a bit over 120 years old. The first electrocardiogram was actually published in *The Journal of Physiology* in 1887 by a Paris-born medical scientist raised and medically trained in Scotland. Although he completed medical school in Aberdeen in Scotland, he took up a teaching position at St Mary's Hospital in London, and it was there that he published his invention of "the electrocardiogram." Augustus Waller actually used to demonstrate that he could detect electrical heart signals from a living being by placing two paws of one of his bulldogs, Jimmy, into pots of salt water. The salty water conducted the electrical signal to a nearby printer. After one such demonstration at the Royal Society in London, the British member of Parliament Viscount Herbert Gladstone was asked in Parliament about possible "cruelty" to animals. His reply made generous use of the caustic British humor so readily deployed in Westminster, the seat of Parliament: "I understand the dog stood for some time in water, to which sodium chloride had been added, or, in other words, a little common salt. If my honourable Friend has ever paddled in the sea, he will understand the sensation. The dog—a finely developed bulldog—was neither tied nor muzzled. He wore a leather collar ornamented with brass studs. Had the experiment been painful, the pain no doubt would have been immediately felt by those nearest the dog." The viscount doubtless delivered this rebuke to uproarious laughter. The electrocardiogram itself, or ECG, is little changed since. It remains a sensitive test for a whole range of heart abnormalities. Every patient who visits a cardiology clinic will receive an ECG that often, to this day, prints slowly and suspensefully from the side of the machine on paper with a pink grid on it. As Steve's ECG slowly *brrrred* out of the machine, it was with a large sigh of relief that I noted it was normal.

Next was an ultrasound of his heart and an exercise test. High-frequency ultrasound waves have been used to reveal underlying organs and, most typically, fetuses since the middle of the twentieth century. It is a very powerful technology to look at size, structure, and blood flow through the heart in real time. Such ultrasounds of the heart are called *echocardiograms* (usually shortened to *echos*). Sometimes, we overlay color on the video to represent the speed and direction of the blood flow. To watch the heart move like this is to be witness to something truly awe inspiring. It is one of the most mesmerizing sights you will ever see. Four chambers, independent but intimately linked, singing their own four-part harmony, repeated three billion times in one lifetime.

Since this was a last-minute, out-of-hours study, we begged a favor of our amazing lead echo-sonographer, Josie Vinoya. Josie performed her study in our exercise lab on the second floor of the hospital, which was, at the time, a bare, beige-walled room about twelve feet by twelve feet in which was a treadmill, a medical examination table, a metabolic cart housing equipment to measure oxygen and carbon dioxide, a sink, a doctor, and a nurse. And into that tiny room, we wheeled another large object: the ultrasound machine. Applying warmed gel to the probe and placing it on Steve's chest, Josie produced her characteristically crystal-clear images while molding herself into the exact three-dimensional shape to fit in the tiny space that was left by these large objects. At times, she would turn up the volume, and we would hear the blood in Steve's heart whoosh through a valve. We all looked at the screen, transfixed. Another relief: Steve's heart contracted well at rest. So it was time to exercise. He chose to bike rather than run on the treadmill. We call these *max* tests, as in *maximum,* because over the course of ten to fifteen minutes, we gradually increase the resistance until the patient can't do any more (doctors are really fun friends to have). It's exhausting, and if you've done it right, your patient will be breathing hard and feeling a bit nauseated at the end. But in reality, we don't need to push people to the point where their muscles produce a lot of excess acid (that's what causes some of the pain and a lot of the nausea). We need to push just far enough that we can estimate the maximal blood output of the heart. After sixteen minutes and nine seconds, 450 watts of power generated by his straining legs, more than a few drops of sweat, and with his heart clocking 191 beats per minute, Steve stopped, exhausted. He had passed the test with flying colors. His maximal oxygen consumption (VO2max—our measure of fitness) was 49.6 milliliters of oxygen per kilogram of body mass per minute, which is 145 percent of what would be predicted for someone his height, weight, and sex. Even better, at all times, the heart walls moved symmetrically, and there was no bad rhythm. In short, Steve's "engine" was currently performing very well indeed.

Next, we turned to his blood results. We had ordered standard blood tests much as your doctor might do in an annual checkup. His liver and kidneys were working well. The electrolytes in his blood were in a good range. But when we examined his cholesterol results, we found something quite concerning.

To understand what concerned us, we need to talk about bubbles—specifically, soap bubbles. When you wash a dirty pan, you probably notice that the oil in the pan, which is basically fat, pushes the water away.

In science terms, it's *hydrophobic.* That's where dish soap comes in. Dish soap, or detergent, is actually something called *anionic surfactant* mixed with a chemical that helps it stay liquid. For most dishes, you probably don't need it—clean water would likely be just as good. But for dishes that are coated in oil, you need something to break up and disperse the oil, and that's detergent. So what does this have to do with measuring cholesterol in the human body? When we measure blood cholesterol, what we are actually measuring is the size and number of fat particles in the blood. But just like those dirty dishes, since fat and water don't mix, after fat is absorbed through our gut into our bloodstream, it is assembled into little packages called *lipoproteins.* The packaging is needed so that fat molecules can travel in a liquid, specifically blood, that is mostly water. The particles are classified by their size (more correctly, their density), and different sizes of lipoproteins do different things in the body. You might have heard of "good" or "bad" cholesterol. Bad cholesterol is a measure of the lower-density lipoprotein particles and is known as LDL (for *low-density lipoprotein*), while high-density lipoprotein, or HDL, is known as good cholesterol. Steve's good cholesterol, which you hope to be high, was 48 mg/dl (1.2 mmol/l)—not high, but not too bad. When we looked at his bad cholesterol, however, which would optimally be below 100 mg/dl (2.5 mmol/l), we found it was quite far outside the desirable range, 156 mg/dl (4.0 mmol/l). For people who have had heart attacks, we usually try to get that number below 70 mg/dl, less than half his level. Even more concerning was another more colorfully named lipoprotein: Lp(a)—pronounced "L P little a." The normal level for Lp(a) is less than 30 mg/dl, with a desirable level being under 15. The high-risk category starts at 50. Steve's Lp(a) was 114 mg/dl, almost four times the upper limit of normal. In the world of rare and severe disease, we pay attention when numbers are just two times the upper limit of normal. This was *way* outside of that. Could this be the origin of his family history of heart and vascular disease? Could it be what caused his cousin's son's death? Should we be aggressively treating that cholesterol despite the fact he himself had never had a heart attack?

It was time to turn to the genome to find out.

Our analysis teams took to their tasks with abundant energy, fueled by the excitement of the unknown. Each team came up with their own three-pronged effort: first, optimizing the resources currently available; second, developing new computational approaches; and third, applying those optimized resources to Steve's genome. Steve became known to us through his code name: Patient Zero.

There were lots of little things to learn. For example, the computer code that detected genetic variants was programmed to ignore anything found in the reference sequence (the sequence from the Human Genome Project). That makes sense if this were truly a healthy "reference" —an idealized version of every gene. But because the reference sequence was derived from real people, it turned out to contain quite a lot of variants associated with disease. So we needed to rewrite that code. There were also bigger challenges—like how to organize all known genetic knowledge about human disease in a way that was searchable and deployable to one genome. Our rare disease team was looking through Steve's DNA both for variants already known to be associated with disease and for additional genetic disruptions of a type likely to cause large disruptions in gene function, even if they hadn't been seen before. There was the question, not resolved to this day, of how to "add up" the effect of multiple common variants, each potentially contributing just a small amount to the overall disease presentation. Pharmacogenomics had so far mostly focused on commonly found variants in specific genes known to be important for metabolizing drugs. Now that sequencing was also revealing rare variants of potentially greater effect for those genes, how should we incorporate those? And then, we had the

98 percent of the genome that didn't code for any genes (the part of the genome formerly known as "junk" DNA). What, if anything, could we say about that? Finally, there was the question of how you might visualize all of this in the context of a medical encounter, making it digestible both to doctor and patient.

We worked for about six months flat out. We worked in small groups and in large groups; at home and at Stanford; during the day and, mostly, at night (we are computer people, after all). It is the team meetings I remember most vividly. The dynamic had some urgency, not just because of the common cause binding everyone together, not just for the excitement of doing this for the first time, but also because of the unusual situation of focusing—as deeply as any of us ever had—on the total genetic risk of just one person: a person who was actually *in the room.*

After six months and hundreds of hours of work from a team of about thirty humans and many more computers, we were ready to return the results. We gathered in the Clark building, where it had all started, in a room just past Steve's office. It was a conference room with a lot of light, a few long tables organized in a U shape facing a screen, and a bunch of plastic chairs on wheels. About fifteen representatives of the various teams sat around the room. Kelly from the ethics team was there, and near the front, slouched into one of the red plastic chairs (somehow, he had a way of always looking relaxed), was Steve. We were a bit less relaxed, unsure if he was ready for what we were about to offer up, but still, it all felt kind of futuristic and a little wondrous.

So what did we find? Our most important finding was a negative one, in a good way. We had developed an approach to look at important gene disruptions throughout the entire genome. We also "fact-checked" a number of gene variants thought to cause disease. Fortunately, we found neither evidence of a rare heart disease in Steve's heart nor genetic variants that could be confidently associated with such diseases in his genome. The variant I had seen on Steve's screen in his office that first day turned out to be mis-annotated in the database. This was a huge relief, since his family history of heart disease, including a teenager's sudden cardiac death, was the major reason for embarking on the analysis to begin with.

However, we did find significant risk for a different form of heart disease: heart attack. A heart attack is caused when a cholesterol plaque inside a coronary artery ruptures and a clot forms, blocking the artery, depriving the muscle downstream of blood and oxygen. Diseases like those underlying heart attack are called *complex* because there is rarely a single, all-powerful genetic variant that causes the disease but rather a complex combination of a large number of variants interacting with environmental factors like diet and exercise. Consistent with his blood work, we found that Steve carried a significant number of these variants that had been associated with high cholesterol and heart attacks. His genetic risk for obesity and diabetes, too, appeared larger than average. We still had no way of knowing how these variants would add up to cause risk (formalized and rigorously developed genetic risk scores were still years away). Yet graduate student Alex Morgan and postdoc Rong Chen from Atul's group found an elegant way of illustrating the risks we could so far define, along with our confidence in the estimate. Graduate student Joel Dudley, also from Atul's team, mined the literature for associations between genetic risk and environmental variables, such as diet, exercise, and environmental toxins, and displayed them together in a novel way, using a circular plot with environmental factors displayed as words with a size relative to their effect and arrows connecting genetic and environmental factors that were known to impact each other.

The most abnormal blood test had been Steve's Lp(a). Could the genome explain that? We knew his Lp(a) was high now, but had it been high his whole life? In other words, did he have a genetic basis for this? Fortunately, a study from colleagues at the University of Oxford had just been published. Martin Farrall and Hugh Watkins described, in *The New England Journal of Medicine*, the association of common variants in the gene for lipoprotein "little a" (*LPA*) with the size of the fat particle produced, the lipoprotein. Specifically, common variants seemed to be a useful flag for a curious phenomenon that affects certain genes. For *LPA,* the size of the gene in the genome actually varies substantially from person to person. In this case, a region of 342 base pairs (called, rather festively, a *Kringle repeat*) can be duplicated more than fifty times. In other words, some people can have an extra 17,000 DNA letters in that gene. You might think that was a bad thing (it certainly sounds like it should be) and *usually* long expanding repeats in important genes are bad, but as it turns out, in this case, the larger the gene, the larger the resulting protein, and the *lower* the circulating Lp(a)! No one is exactly sure why. The theory goes that it takes longer to build a bigger protein, and the liver can't release the protein into the blood until it is ready, which leads to a lower circulating level. Unfortunately, of course, Steve's gene was on the smaller side, with fewer Kringle repeats, and the protein more readily released into the circulation. This contributed to his high blood Lp(a) level and meant that he carried a likely 400 percent or higher risk of heart attack compared with the general population.

So what could we do about this? Steve wasn't on any medication, but clearly, we were thinking he should be. We looked to national guidelines. These, at the time, went by the catchy name ATPIII, which stands for *Adult Treatment Panel III.* To give you an idea of how comprehensive these guidelines were, they had twenty-seven authors and twenty-three reviewers, and the *executive summary* was forty pages long! Fortunately, the recommendations of this learned panel were distilled into a decision tree on one half page where you input your patient's details at each of six steps, and at the end, you get your recommendation. So we entered all Steve's numbers to the decision tree: age, sex, LDL, family history, and more. We followed the tree branches down and cast our eyes expectedly at the bottom right-hand corner of the page for our answer. The expert consensus for Steve was, "Sorry, we can't help you." It literally suggested, in this case, that we should use our "clinical judgment." The reason for the hedge was that Steve was on the border for many of the decisions being made in the tree. He wasn't quite old enough at forty-two for age to be a significant risk (that started in this guideline at forty-five). His LDL was 156, but numbers bigger than 159 were considered high in this document. His HDL was 48, but in this document, you didn't get "risk points" for that unless it was below 40. In the end, the report's recommendations were entirely unhelpful for Steve, leaving us to ponder statements like "drug therapy optional" in our attempt to use our "clinical judgment." What the guidelines were really asking us was, "Do you, as the physician and patient, have any other data that could help inform this decision?" Like a genome, perhaps?

In Steve's genome, we had found genetic risk for coronary artery disease, obesity, and diabetes through the combined effect of multiple common variants. We found additional risk via the small size of his Kringle-poor *LPA* gene. Powerful cholesterol-lowering drugs called *statins* were known to reduce LDL and save lives. Could the genome now help us determine how likely he was to respond to this type of medication? Our pharmacogenomics team had built a new pipeline and created a method to produce personalized prescribing recommendations from a personalized genome. As it turned out, there was a genetic variant known to be associated with a good response to statin therapy. If Steve had that variant in his genome, it would increase his chances of responding to the drug. But what about side effects? If the future were to include truly personalized "genomic" medicine, we should be able to help patients avoid medications that would cause side effects in *them*. Well, there was also a genetic variant known to be associated with the muscle aches and pains that can accompany statin use in up to 10 percent of people. When we looked in Steve's genome, it was good news. We found he *did* have the variant predicting benefit from statins, and he did *not* have the variant predicting side effects. Finally, we took note of another study where the investigators had focused on patients with high Lp(a) and observed a decreased risk of heart attacks in those patients with the use of aspirin.

So we took stock. Patients at risk for heart disease globally: billions. Patients sitting in a clinic asking their doctor about risk for heart disease with their whole genome sequenced: one. To have this much information available to us to make a medical decision was, on the one hand, mind-boggling, yet the judgment calls we now had to make were, in contrast, quite ordinary. We live in a messy world, and as doctors, we make personalized decisions based on imperfect data every day. The patient in front of us rarely fits the exact description of the patients entered into the studies drug companies use to establish safety and effectiveness of their drugs.

We decided, in the end, to recommend a statin to Steve. While that action, that decision, was at once simple, trivial, and repeated thousands of times daily by doctors around the world, it struck us, on another level, as quite profound. Doctors at that point just didn't have access to their patients' whole genomes to help make medical decisions. More importantly, few doctors had the benefit of teams of scientists, bioinformaticians, ethicists, clinicians, geneticists, and more.

So we wrapped up our analyses and got ready to tell the world, or at least that small corner of the world who would be interested. When you have something new that seems exciting, you sometimes call an editor at one of a handful of very widely read journals. I had grown up reading *The Lancet,* a London-based publication dating from 1846. It had a strong worldwide readership, and so I was excited to see if there was interest there. I spoke to one of the editors, Stuart Spencer, and he thought the story was intriguing enough to send for review but, of course, there were no guarantees. I submitted the paper, including the names and emails of all thirty-one authors into the journal's online submission portal, and we shipped our manuscript off, preparing for a few tense weeks of contemplating what our reviewers might think. These reviewers are learned peers, whose names are usually kept secret, who assess the work and provide feedback as well as a yes-or-no recommendation as to whether the paper should be published.

While we waited, we started to think about what might come next. Clearly, we had to start to prepare for a world where genomic information on our patients was much more available. We had to plan for how to deal with large-scale data in a medical world accustomed to kidney tests comprised of just one number or cholesterol tests comprised of just four. What changes would need to be made to deal with thousands of patients with billions of data points each?

When the reviews came back, as usual, our reviewers pointed out some important ways we could improve the paper, disagreed with us on some things (and with each other), but, in the end, appeared enthusiastic about what we were describing. Based on their advice, Stuart asked us to revise the paper along the lines they recommended, which usually means you are on the road to publication.

In the paper, we had struggled with how to deal with the consent and ethics questions that our team had wrestled with. These seemed so fundamental, yet also at odds with the terse style of a clinical/technical data paper. Eventually, after failing to condense all our thoughts into one paragraph of the main paper, we decided to ask the editor if he would consider an accompanying short commentary on just that aspect. In the end, as well as a nice perspective commentary from our colleague Nilesh Samani in the UK, there was an opportunity for us to frame an approach to genome consent and the ethical questions raised by an "everything" test as an accompanying paper. The framework, both technically and ethically, represented the foundation on which we built our approach to genome sequencing for clinical medicine, and it is still with us today. The papers appeared in the May 1, 2010, issue of *The Lancet,* a little over a year after we started work. The editors chose one of the lines from our paper for the front cover: "As whole-genome sequencing becomes increasingly widespread, availability of genomic information will no longer be the limiting factor in the application of genetics to clinical medicine."

I am not quite sure what we expected when the paper came out. We were certainly excited about the work, and I knew people were fascinated by genomes, but we were not really ready for what happened. For days before and immediately after the paper came out, we were taking phone calls from morning to night, talking to journalists all around the world. Steve and I did a "doctor-patient" interview for National Public Radio's *Morning Edition* with Richard Knox. I did an interview for BBC radio. A Japanese TV team came out for several days to film a documentary. A Turkish newspaper made a wonderful anatomical image overlaying our findings relating to Steve's various organ systems on this body. In fact, the Google News trends line provided one of the more humorous perspectives on all of this. It showed a time line of interest in the story, essentially the number of new stories coming out about it—a graph I still sometimes show in presentations. Since *The Lancet* crossed the wire shortly after midnight, the first stories came out post embargo in the late evening California time and, at that time, the graph showed a spike. Then, when the U.S. stories came out the next morning, there was another spike—we were celebrities! Then, almost immediately, the graph spiraled downward rapidly (a cliff not unlike the graph showing the drop in the cost of sequencing) such that by 4:00 p.m. that day, there were no more stories. None. Nada. Zero. It was late afternoon, and our time was up. We were, once again, nobodies! Even scientific fame, it would appear, is fleeting.

Although the main burst of news coverage was short, we were humbled by a lot of continuing interest in our approach to interpreting the human genome. Kevin Davies, the founding editor of the journal *Nature Genetics* and the author of a book titled *The $1,000 Genome,* was kind enough to highlight our work at the International Congress of Human Genetics in Montreal. We were in some very esteemed company as he recapped the history of the sequencing of the human genome and beyond in the opening plenary prior to a panel featuring Jim Watson, the Nobel Prize winner who helped elucidate the structure of DNA (the same Jim Watson whose genome had been sequenced by the Baylor team). Shortly afterward, the National Human Genome Research Institute was putting together a genome exhibit for the Smithsonian, and the project was selected to be featured. For several years following, at random intervals, Steve and I would get calls from friends and family as they visited the exhibit, only to be surprised to be met by familiar faces.

I think what captured people's imagination was the hint of something tantalizingly futuristic: the idea that deep insights may be revealed when we start routinely decoding the genomes of individual people as part of medicine. Most people have an intuitive sense of what personalized medicine is or should be. They have personalized their diets, their exercise regimens, their wardrobes, their kitchens, their gardens, and their cars. Indeed, as a species, we are marked by our ability to express ourselves as individuals. People seemed primed to imagine having their doctor acknowledge their individuality at least as much as their hairdresser. Yet beyond this, the genome holds a mysterious place in our culture. In sequencing the first human genomes, some believed humans were reading "God's language" for the first time. Now, here in the genome of our Patient Zero was a glimpse of what might be to come: a future where we could predict disease more powerfully, describe disease more accurately, and prescribe drugs more precisely.

And with regard to that, we were only just getting started.

## 3

# Once Removed

*When you part from your friend, you grieve not;*
*For that which you love most in him may be clearer in his absence,*
*as the mountain to the climber is clearer from the plain.*

—KHALIL GIBRAN, *The Prophet*

They had all been watching the TV show *Lost* one evening—the fourth season, the one where a freighter appears off the coast of the island and the castaways fight about who might actually get to go home. It was a Thursday night in mid-February, and the hour was late. The family was heading off to bed. Richie, the oldest child of Marilyn and Rich, was completing a co-op program at Drexel University in Philadelphia, and his internship was near home, so he had moved back in with his parents for the winter to enjoy some home comforts. He was a naturally athletic and talented student. That night as they headed to bed, his dad asked him to help his little sister with her preparation for an upcoming advanced practice test. "My daughter and I are so similar," Rich told me later. Richie was "like his mother—he could always straighten things out." That night, as he had many times before, Richie went to go help his little sister prepare for her test.

The family was up and out early on Friday morning while Richie slept. He had been feeling a bit cold when he got up, and before leaving, his mom bundled him back to bed with the cat. When she got home later that afternoon, she remembered wondering why her son's car was in the driveway when he was supposed to be at work. She walked upstairs to his bedroom and opened the door. Her scream brought her daughter running. Richie was lying motionless on the floor. When his dad got home, he sprinted up the stairs, but it was clear from his pallor that his son had been gone for some time.

When someone dies outside of the hospital, we physicians experience the death mainly through cold, clinical words on a page: the eerie aura of an autopsy report. You are discomforted, peering into an intensely private past moment, usually one that, in death, changed the lives of the living forever. Autopsy reports of sudden death start with a summary of the circumstances surrounding the discovery of the body. This generally comes from the police, who are often called when the death is sudden and unexplained. It includes a description of the way the deceased was found: their clothes, their body perhaps in an awkward position, sometimes covered in blood. Then, abruptly, the report switches from the place of death—a home recently filled with warmth and happy memories now punctured by the icy barb of grief—to the autopsy suite. The deceased is now laid straight out on a table, unclothed: as in birth, as in death. Their features are described matter-of-factly: height, weight, eye color, hair length, and identifying moles or tattoos (a list of features no one who would recognize them wants to read). The craft of opening the body, examining the organs, generating hypotheses about the cause of death has changed little in centuries.

In sudden unexplained death, the focus is the heart. The pathologist weighs and measures it, notes its size, details the thickness of its walls, opens its arteries and veins. Then tiny pieces of heart tissue are placed in paraffin blocks and razor-thin slices cut to be examined under a microscope. Sometimes, muscle fibers are abnormally arranged; sometimes, the muscle is replaced by fat. Sometimes, an artery is found to be completely blocked by a fresh blood clot. Each of these is a clue pointing toward the most common identifiable causes of sudden death at autopsy. More often than not, however, there is no smoking gun. In fact, in sudden unexplained death in the young, the most common autopsy finding is no finding at all.

"Presumed cardiac death, cause unknown" was the official result of Richie's autopsy. For a parent to bury a child is an unfathomable hardship. To bury a child while having no idea why they died and no handle on how to protect the remaining children from the same fate is unconscionable. So Rich began a singular quest. It dominated his every minute, his every thought, the late hours, and the early hours. Driven by the fury of grief, he would do everything in his power to protect his family.

He turned to the internet. He called friends, neighbors, friends of friends. He got connected to experts at Mount Sinai Hospital in New York City, at the Mayo Clinic in Rochester, Minnesota, at patient organizations that serve families impacted by sudden death. As the months went by, he talked to literally anyone who might help. He also reached out to his broader family. He had a cousin in California who was a scientist, a renowned innovator. That cousin, he'd heard, had sequenced his own genome. So one day, Rich Quake picked up the phone and called his cousin Steve to ask whether genome sequencing might help them work out what happened to Richie, Steve's first cousin once removed. And as we began to finalize the analysis of Steve's genome, our minds turned to the idea of using those same tools to see if we could solve the mystery of Richie's death.

. . .

The "we" in this case included a new addition to our team. Frederick "Rick" Dewey came highly recommended by two people whose opinions I have learned to value greatly: Victor Froelicher and my wife, Fiona. Vic Froelicher was the reason I first came to California from Scotland in 2002. Vic is tall and lean and approaches life with a manic, unbounded energy. Vic was a pioneer in the use of exercise testing to diagnose heart disease, starting in the 1960s in his air force days, and had authored hundreds of papers and several textbooks when I first met him in the mid-1990s. As a medical student, when I wrote a letter to Vic to request a summer internship, he wrote back a note suggesting we continue to communicate via "email." His was only the second email address I ever had (those were the days in Scotland when you had to walk a quarter of a mile to the computer science department to send and receive email). So when Vic called, I answered, and when he told me I should take on this trainee because he was one of the smartest trainees he'd ever met, I was excited. Rick Dewey had studied physics and chemistry at Harvard, after once beating a swimmer known as Michael Phelps in a junior championship swim meet. Having then swum for Harvard, he moved west for medical school at Stanford and joined the local rowing club, Bair Island Aquatic Center in Redwood City. Through the rowing club, he also met my wife, who provided a strong character reference, mentioning that the only thing as outsize as his talent was his modesty. So Rick joined our lab. I was on the lookout for a project to stretch him, and if there was a project to stretch someone computationally talented but with no real background in genetics, then extending the tool kit we built for Steve Quake's genome to do a comprehensive whole genome "molecular autopsy" was probably it. I talked to Rick about partnering with Matt Wheeler, who had led the analysis of Steve Quake's genome in our lab, to look at Richie's genome. He was all in.

We first set about acquiring the tissue block containing a tiny piece of Richie's heart from which we could obtain his DNA. Coroners usually keep these for many years and are very happy to share them, with the family's permission. Rich had made sure everything that could possibly be relevant to understanding Richie's death was saved, and in fact, a small handful of genes known to be associated with sudden death in young people had already been sequenced, with nothing found. After acquiring the block of tissue, we dissected a small piece and isolated DNA from it. We then sequenced the whole genome using an updated version of the Helicos sequencing technology used for Steve's genome.

Despite the fact that the DNA was derived from a postmortem heart sample, we were able to generate more than double the quantity of sequence data from Richie's genome as compared to Steve's, partly because of technology advancements. This was about a year later, and we had also updated the custom software we had used for Steve's genome, but when it came to deciding which genes to look closely at for a possible cause of death, we needed a new approach. What we needed, in effect, was a list of every gene that could possibly contribute to sudden cardiac death. Such a list did not, to our knowledge, exist. So we set about making it. This involved some new-school and some old-school techniques. We started with all the genes known to cause heart muscle disease and then expanded it to all the genes and proteins known to be active in the human heart. We also used some of the approaches we had developed to try to understand the connections between genes—in essence, using a social network approach to find genes that were "friends" of important genes. We knew from other work that "well-connected" genes were important in biology and that connectedness could be a more powerful predictor of importance than how active a gene was (sort of like guessing about someone's influence from a list of their friends rather than how loudly they talk). Then, we went old-school: we went back through every classic textbook we could find and pulled every protein and every process. Rick mapped these to genes, and finally, we had our list. Now it was time to examine those genes in Richie's genome.

We already knew that Richie had no mutations in genes already well known to cause sudden cardiac death, because Rich had made sure standard clinical genetic testing had been done. Also, no heart artery blockages had been seen on the autopsy, suggesting cholesterol genes were not the problem. So Rick focused in on the lesser-known genes, in particular, the genes that control the electrical excitability of the heart. These are mostly specialized channels that shuttle sodium, potassium, or calcium in, out, and around the heart cell. Two genes in particular had variants in them that looked suspicious. One coded for a channel that shuttled potassium, while another shuttled calcium. Could they be working in concert to cause electrical irritability? It seemed quite plausible. An experiment where we inserted the mutated channel into cells even suggested a significant abnormality in the potassium channel's functioning. But then, as so often happens in science, our beautiful hypothesis was shattered by an ugly truth. Rick was looking more deeply into this variant and found that the individual stretches of DNA in this region from Richie's genome were not being mapped correctly to the reference genome. In effect, the jigsaw pieces had been placed in the wrong position by the computer program. In trying to explain why, we uncovered a challenge that would come to haunt us for many years: the potassium channel gene had a close relative in the genome, a pseudogene. Probably as many as half of our genes have pseudogenes that have arisen at some point in our long biological history because a gene has been duplicated or a version of it has been reinserted into the genome. They are called *pseudogenes* because they are mutated compared to the "real" gene and have lost some functionality. Usually, for example, this means their RNA "message" does not lead to a protein being made. When we compared the DNA from Richie's genome, not just to the human reference sequence but also to a few other anonymous reference genomes that had just recently been sequenced, we found the variant was far too common to be causing a phenomenon as rare as sudden cardiac death. This was disappointing, but we had learned a very important lesson about the complexity of the genome and the challenges of re-creating one long sequence from a series of smaller jigsaw pieces.

So the potassium channel variant was out, but our other candidate, the calcium channel variant, was still in play. It became our major focus. We knew other calcium channel genes could cause cardiomyopathy and sudden death, and since this gene was important for cardiac excitability and the variant in this gene had never been seen before, it was a good candidate for causing something as rare and devastating as sudden cardiac death.

Rick presented our results at the American College of Cardiology meeting in 2011 in front of a large group of several hundred cardiologists. I was very proud of him, as this was one of his first presentations at a national meeting. Later, one of our Stanford writers, Krista Conger, wrote a beautiful piece on Richie and his family for a special edition of the Stanford magazine.

The idea of postmortem genetic testing to help establish a cause of sudden death has come a long way since the first publication in 1999. Michael Ackerman at the Mayo Clinic identified the molecular cause of an electrical disease of the heart (long QT syndrome) from sequencing some parts of four genes from a patient's postmortem tissue. Although whole genome sequencing postmortem remains unusual, in 2016, a group led by Christopher Semsarian at the University of Sydney and Jon Skinner at the University of Auckland showed that among young people who died suddenly in Australia and New Zealand, a relevant genetic variant could be found in almost a third of cases. Surprisingly, even in those under the age of thirty-five, coronary artery disease—what we used to think of as a disease of middle age and beyond—was a common cause of death. But as in most studies in this field, unexplained sudden death was the most common finding of all, the final result in 40 percent of cases.

The promise of postmortem genetic testing goes beyond providing an "answer" to the question of what caused the death: it can also help those left behind. Since many sudden deaths in young people turn out to have a genetic cause, family members are often unknowingly at risk. Using the information from the genetic test, we can test other family members and identify who is or is not at risk. Then we can protect them using medications or lifestyle changes or, in some cases, implanting a device under the skin that can deliver a lifesaving shock to the heart in the event of a dangerous heart rhythm.

In the United States, we continue to battle the paradox of a health care system based on individual subscriber insurance. Since the deceased don't have health insurance, the system lacks a mechanism for paying for the test. The payors have no legal responsibility, though it is hard not to argue for a moral responsibility. The system also ignores that genetic testing of the dead to protect the living clearly has the potential to avert emotional anguish, save lives, and save money. Despite this, most of the postmortem genetic testing we carry out today is paid for out of pocket by the family of the deceased or by us through charitable donations to our Center for Inherited Cardiovascular Disease.

Steve Quake's and Richie Quake's genomes each in their own way represented a glimpse into the future of genomic medicine. On the one hand, using the genome to diagnose severe genetic disease, as we did with Richie, would be adopted quite quickly, at least for the living: the genome or, more commonly, a test where every gene is sequenced (an *exome* representing about 2 percent of the entire genome) is now used routinely for rare, genetic disease. On the other hand, using genetic information prospectively, to help prevent disease or personalize the broader practice of medicine, as we did with Steve, remains a niche practice. Despite dropping prices, few genomes are sequenced solely to guide preventive care.

"You can't do two until you've done one" is a phrase my team is, very probably, tired of hearing from me. D. J. Patil, the USA's first chief data scientist, put it better, and more famously, on White House notepaper: "Prototype for 1x, Build for 10x, Engineer for 100x." I used to give a talk on the Quake genomes entitled "The Dawn of Genomic Medicine." After a year of seeing this talk evolve with new data but an unchanging title, one of my friends politely asked if perhaps the "dawn" had now broken and suggested we might now be approaching at least the "morning coffee break" of genomic medicine. Over the next few chapters, I'll describe how our whole community took the next steps along this exciting path.

I will let you judge for yourself if we ever made it to lunch.

# 4

# Genome Illumination

*My fancies are fireflies. Specks of living light twinkling in the dark.*

—RABINDRANATH TAGORE

*Always do sober what you said you'd do drunk.*
*That will teach you to keep your mouth shut.*

—ERNEST HEMINGWAY

It seems particularly fitting that the British company that was to revolutionize human genome sequencing had its origins in an English pub over pints of warm beer. In the summer of 1997, a series of "beer summits" took place in the Panton Arms in Cambridge, UK. Shankar Balasubramanian and David Klenerman were both junior faculty members in chemistry at Cambridge, and they were joined in the pub by their postdocs Mark Osborne and Colin Barnes. Over pints of the local ale, they discussed how to use fluorescently labeled nucleotides illuminated by laser light to observe the action of the enzyme DNA polymerase as it synthesized a single DNA molecule immobilized on a surface. Here, the key breakthrough that would lead to a technology that would come to dominate the global landscape of next-generation human genome sequencing was born.

Klenerman had a laser system in his lab, which is what led to his initial meeting with Balasubramanian, who was a nucleic acid chemist. They started talking about collaborating and found a common interest in the synthesis of DNA. They discussed ways of tracking its movement and had an inkling this could lead to a faster way of sequencing DNA. Although by their own admission, they were just playing around, they were serious enough to approach the venture capital firm Abingworth to fund a new company with the promise of increasing the throughput of DNA sequencing by ten- to one-hundred-thousand-fold. The company—later named Solexa, from *sol,* meaning "light," and also implying single or solo molecules—set up its first lab just outside of Cambridge. It was a big step up from the university hardware but also, still, a long way from the engine of the genome revolution it was destined to become.

Enter Clive Brown, who was also destined to become both an engine of and, later, a disruptor of the genome revolution, but he wasn't initially too impressed with Solexa. He'd heard of the company and the idea and the people involved sounded "all right," but when he arrived for his job interview, his heart sank. It was the building, you see. "It was basically a shed with a sign on it," he told me, thinking back to his arrival for an interview in 2001. But once inside, he liked the team (the founders had now passed the torch to CEO Nick McCooke, Harold Swerdlow, and John Milton). They also seemed to like him, so he signed on.

Clive had always been into computers. Growing up in Blackburn in northwest England, he learned to program on a Video Genie machine at nine years old, graduating to a Dragon 32 at twelve years. He also got quite interested in DNA in high school and so decided to go to university to learn genetics. Finding himself a little uninspired by the entire university experience ("bored shitless" were his exact words), he took an internship at a pharmaceutical company, which he found more stimulating. After finishing his degree, he moved to Glasgow, Scotland, to consider a Ph.D. but remained ambivalent enough about academia that he diverted himself into a master's in computer science. This took him to the renowned Sanger Centre in Cambridge, a center which, under his mentors Richard Durbin and David Bentley, had completed almost a third of the sequencing for the Human Genome Project, making it the largest single contributor. "So that sort of made me a bioinformatician," he reflected, with trademark understatement (at least where his personal achievements are concerned). At Solexa, he was excited by what the technology could offer, but at least initially, he remained skeptical about the actual machine. "I mean, literally nothing worked," he remembers of those early days in the new Solexa lab. "They'd been able to emulate what Shankar and David had done [in the university]. Which is basically get some fluorescence on bases, get some DNA, and see blobs."

The idea basically was to watch DNA being made. The nucleotide building blocks, A, T, G, and C, were supplied to the DNA polymerase copying machine, which was working on a piece of DNA stuck down to a glass slide. The building blocks were adapted to have mini fluorescent light bulbs attached. The sequencing machine then used a laser light to turn on the light bulbs and a microscope to "read" which letter was incorporated at each site into the DNA template. There could be millions of sites on a slide, something that could greatly scale up sequencing. The challenge was that the fluorescence from a single molecule was not very bright. They needed a way to turn up the power of the light source.

The company's ambition was significant, however: they aspired to sequence human genomes. For that, they were going to have to think bigger and, in particular, look toward the American market—they started looking for a new CEO.

John West had heard DNA was going to be a big deal. Born in Detroit, Michigan, where his dad worked for the Ford Motor Company, he completed his undergraduate and master's degrees in engineering at MIT, building robotic systems for assembling semiconductors. He then joined a company called Bio-Rad to lead their research and development team. Bio-Rad had automated the reading of the photographic films that were used in early Sanger sequencing to identify the order of DNA letters from the patterns made by radioactive DNA as it was pushed down a big slab of gel by electricity. Even then, he remembers, distant possibility though it was, the idea of sequencing a human genome. "I remember this was very aspirational," he told me. Enjoying the business side of engineering innovation, John completed an MBA at the Wharton School of business and then joined, and later led, Princeton Instruments, a company making cameras that could pick up very low levels of light. One of the first customers for that camera, in 1998, was a small British company named Solexa.

By the time a member of Solexa's board reached out to John in 2004, he was with Applied Biosystems, a company specializing in automated Sanger DNA sequencing using fluorescent technology—and whose low-throughput technology was the workhorse of the Human Genome Project. The voice at the other end of the phone asked him if he had heard of a company called Solexa and if he might be interested in a job as its CEO. He understood the fluorescence-sequencing technology invented by Solexa but was unclear how they could produce sufficiently reliable fluorescence signals from single DNA molecules, so he turned the offer down. It wouldn't require a large increase in the fluorescent signal to achieve reliable imaging, he figured, perhaps only one or two orders of magnitude, but Solexa simply didn't have the goods.

A few months later, however, he learned that Solexa had licensed technology from a company called Manteia for creating DNA clusters—small "islands" of DNA on which around one thousand identical DNA molecules could be simultaneously "grown." This caught John's attention. In comparison to a single DNA molecule, the cluster technology could increase a fluorescent signal by three orders of magnitude. In fact, John had previously encouraged Applied Biosystems to buy that same technology, suspecting it could be transformative to sequencing, but the leadership hadn't been interested. When he saw that this small British biotech had acquired the rights, he called back their chairman. "Do you still need a CEO?" he asked.

Over the next couple of years, as John West came on board, the Solexa team completed a major milestone: they sequenced the genome of the tiny organism PhiX174, the same one whose 5,386-nucleotide genome Frederick Sanger sequenced in 1977. The fact that Clive announced the successful sequencing of PhiX174 not in a scientific paper but in a company email in February 2005 was testament to the real ambitions of the company: launching a competitive product and sequencing human genomes.

The path from a single-stranded DNA virus with 5,386 bases to a human genome with 6 billion looked long, but the company was expanding. In 2005, John and the management team at Solexa, including the former CEO Nick McCooke, completed a reverse merger with a struggling RNA-sequencing company called Lynx Therapeutics based in Hayward, California. Solexa, a private company, took over Lynx, a public company, and with it the NASDAQ listing, the customer base, and the distribution system for commercial sequencing machines. With two distinct sequencing technologies now in the merged company, a major strategic decision had to be made with respect to which technology the new company should adopt. Naturally, the Californians favored their own technology, and some even foresaw a winding down of the UK operation. But John, together with Clive Brown and others, felt the potential of cluster sequencing was simply greater than the Lynx approach, so the decision was made: the Solexa technology was in. The team moved rapidly toward launching the first commercial system and the Genome Analysis System, the first sequencing machine to feature cluster-based DNA sequencing by synthesis, shipped in June 2006.

There was intense interest in the launch. The early customers understood that the next-generation sequencing machine was barely more than a prototype and, according to almost everyone on record, often failed. But the orders kept coming because scientists could see the potential of the technology to massively scale sequencing capacity. Instead of sequencing hundreds of DNA molecules, this machine could read the sequences of millions of different molecules simultaneously. The surrounding buzz soon attracted the attention of one of the leading companies in the genomics space, Illumina, and of their CEO, Jay Flatley. Illumina, based in San Diego, California, was at the time the industry's leading manufacturer of "oligos"—(oligonucleotides—short pieces of DNA that formed the basis of microarrays, a technology that could measure the expression level of thousands of genes on one glass slide). Yet they were interested in expanding this dominance, saw a future in sequencing, and were looking to acquire technology that might allow that. Before Flatley could pick up the phone, however, John West came to him.

At Solexa, the team appreciated that, even with their efficient new machine, the cost for sequencing a human genome was going to be high. The first instrument could sequence a billion bases, which seemed extraordinary at the time, yet to accurately report a human DNA sequence with this technology, it was not enough to look at a given spot in the genome just once. Instead, it required sequencing that same territory more than twenty times. To cover a human genome with sufficient depth as to provide confident calling would therefore require nearly one hundred billion bases of sequencing. And since it took three days each time the machine ran, it would take up to a full year for one machine to sequence a human genome while running full-time. That did not seem like a scalable business plan. So West and his team started to think creatively about concentrating on the parts of the genome of maximum interest. What about sequencing just the gene sequences and ignoring the other 98 percent of the genome? Although by this time, it was clear that the non-gene part of the genome was definitely not junk, its function, especially in relation to disease, was less clear. Perhaps they could sequence just the 2 percent of the genome that was made up by genes? Their plan was to synthesize a few hundred thousand oligos and to use them as bait to fish out just the gene regions of interest. That would require much less sequencing and incur much less cost: something that could be completed perhaps in just one run. They needed a supplier for their oligos, so John reached out to Illumina.

At the end of a successful meeting with Illumina's CEO in his San Diego office, at which they agreed on a large order for oligos, John got up to leave. Flatley stopped him. "Is that it? There wasn't anything else you wanted to talk about?" he asked. Soon after, in 2006, Solexa was acquired by Illumina in a stock deal worth $650 million. Today, that stock alone is worth close to $9 billion, and many believe that Illumina's total $40 billion valuation derives mostly from the performance and potential of the sequencing business, to this point, entirely based around the Solexa technology.

Although they were initially thinking of the Solexa acquisition as a mechanism to boost their gene chip business, it didn't take Flatley and Illumina long to recognize the potential in Solexa's technology to open new business opportunities. Shortly after acquisition, and with accelerated research and development, the second-generation Genome Analyzer was launched, delivering reads up to seventy-five base pairs and a throughput of 2.5 billion bases of data per day. This rocketed Illumina to front and center of the global sequencing market. In 2010, an engineering advance allowed clusters to be grown on both surfaces of the glass slide (flow cell), something that dramatically increased the sequence output. In fact, this approach, labeled HiSeq, made possible the simultaneous sequencing of the genomes from five people (or, if sequencing just the genes, one hundred people). This dramatic boost in capacity led to massive uptake of Illumina's machines in the United States, in China, and beyond. The world was waking up to the power of sequencing to revolutionize science and medicine, and the cost for individual exomes and genomes was falling (about a couple of dollars in Ferrari money). In fact, it was falling so fast that Illumina decided to offer "personal" genome sequencing to ten individuals to celebrate the launch of the new machine. Jay Flatley, Illumina's CEO, was first in line. The actress Glenn Close was not far behind. And in that initial group of ten were four people all with the same last name: West.

# 5
# First Family

*We all carry, inside us, people who came before us.*

—LIAM CALLANAN

*The global economy is built on two things:*
*the internal combustion engine and Microsoft Excel.*
*Never forget this.*

—KEVIN HECTOR

John West called me shortly after the paper describing Steve Quake's genome was released. I knew of his connection to Solexa, but we had never met. He introduced himself as an MIT-educated engineer and entrepreneur who had worked on DNA sequencing technology for many years. More important for this discussion, he added, he was a father of two, and he had suffered a pulmonary embolism a few years previously. I had the impression he was asking for medical advice, so I shifted into doctor mode.

A pulmonary embolism is a sometimes serious and sometimes rather stealthy event. A blood clot arises, typically in the legs. Small at first, a clump of tiny little cells together activate a cascade of reactions, turning liquid blood into solid clot, and this clot travels up to the heart and from there into the blood vessels of the lungs, where it transits through smaller and smaller branches. Eventually, it reaches a branch the same size as its diameter. It can't go any farther, so it embeds itself there, blocking the blood flow. The lung tissue served by the branch is starved of blood and oxygen. It dies, causing pain, shortness of breath, and low oxygen levels. The size of the clot determines how far it gets in the branching tree of blood vessels in the lungs: the bigger the clot, the bigger the blood vessel it can clog, and the bigger the problem. In fact, while many of us will die of old age with a few small pulmonary emboli, embedded in tiny little twig-sized arteries in our lungs, likely undetected in life, large pulmonary emboli can cause sudden and instantaneous death.

John's pulmonary embolism thankfully resulted "only" in pain and shortness of breath, and it was successfully treated at the time using the drug warfarin. Warfarin works by stopping the production of specific clotting factors in the liver. It is an effective treatment for blood clots and, in very high doses, for killing rats (it's the main ingredient in some rat poisons). However, in his case, it was not *that* effective, because despite taking the drug exactly as prescribed, he suffered a second pulmonary embolism sometime later. This was worrying and certainly unusual. But why call me? Why not call a blood doctor, since this sounded a whole lot like a problem with blood clotting?

"Oh, I was concerned about how this would affect my kids, so I decided to sequence my family's genomes," he said. Excuse me? Come again? "We've been working on some analysis, and my daughter and I have made some progress, but we hoped you could help us take it further, given the work you performed on Steve Quake's genome."

My jaw nearly hit the floor. My inner voice sought clarification. *So let me get this straight: This is 2010, and you sequenced your whole family? And you and your daughter have been working on the analysis? And you wondered if we could help? And, to clarify, you are talking about your seventeen-year-old daughter, and this is her high school science project?*

Welcome to Silicon Valley.

John and his daughter, Anne, had been working with the *variant call files.* These files, a few megabytes in size, are basically a list of all the places that each person sequenced varies from the reference genome from the Human Genome Project. They were doing this mostly in Microsoft Excel, a remarkable achievement in itself, especially given that Excel had only recently expanded its maximum row capacity from a few hundred thousand rows to the millions required to parse genome files. Anne had begun to map out which parts of the genome she and her brother, Paul, had inherited from each parent.

As John described how far they'd come to me on the phone that day, it was clear that this was an interesting opportunity for us: the first nuclear family to have their genomes sequenced by Illumina, plus an unexplained medical condition. But in the year since the team of teams had attacked Steve Quake's genome, we had all gone back to our regular lives, pushing our various other scientific agendas forward and seeing patients. I had certainly begun conversations to see how we might move some of these approaches into the clinic. But to build genome tools for a whole family? Tools that would account for, and take advantage of, the genome sections shared across a nuclear family? Tools that would compare and contrast risk for disease across four family members? That was a sizable task.

I had a hunch, however, that family-based analysis was going to be really important to the future medical application of genomics. Indeed, it is impossible to view medical genomics in any other light. In genetics clinics, when we first meet a new family, we put together a multigenerational family tree, laying out who is related to whom and who suffered what disease when, to help us decide if the disease is inherited through the mother's side, the father's side, both, or neither. Understanding how the disease is inherited in a family provides great diagnostic power. In our new whole-genome context, a major benefit was going to be the ability to narrow the list of possible disease-causing variants according to who in the family suffers from the disease. If we knew, for example, that a particular disease we are interested in requires both copies of the gene to be affected, we can have the computer narrow our search to fit that scenario. Or if a disease appears to have arisen for the first time in a child, say if the parents are completely normal and a new baby is affected severely, we can have the computer look for a brand-new variant in the child that does not exist in either parent.

With such an intriguing opportunity, how could we refuse? Also, we'd had so much fun the last time around. It must be time to get the band back together.

Over the next few days, I reached out to the team who had analyzed Steve's genome, and when I mentioned a comeback tour, no one hesitated. We were going to make a new album! But clearly, we needed some new players.

A new chair of genetics had recently been appointed at Stanford. Mike Snyder had arrived from Yale and brought with him a very broad set of interests. For example, he was a geneticist who wasn't just interested in genes. He was also interested in the proteins coded for by those genes and the by-products of cellular processes (metabolites) that resulted when those proteins did their work. In fact, he had trained with Ronald Davis, one of Stanford's most renowned inventors. He was also very interested in genome regulation (the science of knowing how genes turn on and off) and had worked on those mechanisms in yeast.

Mike is wiry, intense, self-deprecating, and a self-described workaholic. He appears to bend the rules of space and time, somehow managing to be in multiple places at once. I might be on a conference call, and suddenly his voice will appear, emanating from some far-flung place around the globe where it is 4:00 a.m. and where, having just gotten up to go work out, he thought he'd call in with some helpful thoughts. As soon as he took over as chair, we started to talk about how we might apply the work done on Steve's genome to patients. Mike envisioned a new personalized medicine center at Stanford and had started to build a team of bioinformaticians to plan for a future where thousands of people had their genomes sequenced. Mike was a man on a mission: he wanted to sequence every patient in the hospital. I liked that kind of ambition.

The first person Mike hired as chair was a brilliant young scientist named Carlos Bustamante. Carlos was born in Caracas, Venezuela, but grew up in Miami, where his dad is an infectious diseases doctor and his mom a clinical psychologist. Carlos himself was precocious. As a teenager, he would wear a suit to school during the day and run mathematical models of the New York Stock Exchange at night on his Apple II computer. Clearly, he was destined to be a professor or a financial mogul. After initially enrolling at the University of Miami in a six-year M.D. program, he transferred to Harvard to pursue a budding interest in population genetics—inspired by the writings of the famous scientist Richard Lewontin. No sooner had he arrived in Boston than Richard Lewontin himself took Carlos under his wing. Carlos stayed at Harvard for both a master's and Ph.D., and after a brief postdoctoral stint with the Australian genomics luminary Sir Peter Donnelly at Oxford, he took up a faculty position at Cornell from where he was recruited to Stanford. In person, Carlos appears like he was put on this earth to provide the counter to the word *taciturn.* Brilliant, expansive, entertaining, erudite, and riotously funny, he was going to fit into this team just fine.

So, here we were—the old team plus Carlos and Mike and, as was becoming customary for these genome projects involving engaged early adopters, the West family themselves.

Our first task involved double-checking the one piece of genetic information about the family we already knew. This, we thought, would be simple.

A few years prior, in response to his pulmonary embolism, John's hematologist had tested him for known clotting abnormalities. Among other things, this workup included a genetic test for a single-letter change in the gene for a specific clotting factor, Factor V (five). This variation predisposes someone to overactive clotting. Originally discovered by researchers in Leiden, in the Netherlands, it is known as Factor V Leiden—and John had been found to carry it. The test performed on him used Sanger sequencing and zeroed in on just that one position in his genome. So the first thing we did, now that we had As, Ts, Gs, and Cs for every position in his genome, was to look up that position: two letters (in two copies of the gene, one from his mother, one from his father) out of approximately six billion.

However, as we searched through John's millions of variants, we found a strange thing. Or rather, we didn't find an expected thing. The Factor V variant wasn't there. Was the old test wrong? Or were we reading the genome wrong? What was going on?

To explain the answer, we have to go to Buffalo, New York.

# 6

# Buffalo Buffalo Buffalo

*Oh, give me a home where the buffalo roam*
*Where the deer and the antelope play*
*Where seldom is heard a discouraging word*
*And the skies are not cloudy all day.*

—BREWSTER M. HIGLEY

*I have the right to life, liberty, and chicken wings.*

—MINDY KALING

On a chilly Sunday in March of 1997, an advertisement ran in *The Buffalo News,* making its pitch to this city known for its chicken wings and the grammatical acrobatics facilitated by its name being both a verb and a noun (did you know that buffalo buffalo buffalo?). Local scientists, it said, were looking for twenty volunteers for a large international research effort. "The goal is to decode the human hereditary information (human blueprint) that determines all individual traits inherited from parents," it read. "The outcome of the project will have tremendous impact on future progress of medical science and lead to improved diagnosis and treatment of hereditary diseases." Then farther down, in bold: "No personal information will be maintained or transferred."

The advertisement was placed by genome scientist Pieter de Jong, a principal investigator at Roswell Park Cancer Institute, who was an early pioneer of techniques that would become fundamental to the Department of Energy's moonshot idea: sequencing the human genome. Pieter, originally from the Netherlands, completed his Ph.D. at Utrecht University and cut his teeth exploring the genes of PhiX174, the organism sequenced by Frederick Sanger and by a youthful Solexa. After stints at Albert Einstein College of Medicine and the University of North Carolina, he took a position at Lawrence Livermore National Laboratory in Northern California, where he joined a team thinking about what it would really take to sequence a whole human genome. Like many human endeavors of scope and ambition, it was partly fed by competitive spirit: the Japanese appeared to have seized upon this idea too and were moving rapidly. As the project came together, Pieter was recruited back to the East Coast.

The intent was to avoid having any one volunteer contribute all the DNA for sequencing as this "might raise the interest and curiosity of the public and the press to discover the identity of the donor" as Pieter explained in a paper many years later. Instead, the original plan was to construct a complete human genome using genetic material from ten donors, each contributing 10 percent of the final DNA. As a result, the final sequence would be a patchwork: jumping from one anonymous individual's sequence to another and making those transitions at unknown spots.

There was another unusual feature. A real human's genome is *diploid,* meaning that there are two copies of everything (one from Mom and the other from Dad). The reference genome, however, would be *monoploid,* with just one copy of every human gene and of all the other DNA in between. This single copy of the genome, made up from ten volunteers, would become the standard to which all other genomes would be compared over the following years. This was the Human Genome Project.

The process began with chopping up each volunteer's DNA into pieces two hundred thousand DNA letters long and inserting those pieces into so-called bacterial artificial chromosomes (BACs) that were then introduced into bacteria in the lab dish. As the bacterial colonies grew and divided, each would make many copies of its own short piece of the human genome. Together, the many colonies of bacteria would make up a complete library of the person's DNA. The technique was actually later replaced by *shotgun sequencing*, where shorter DNA fragments were amplified directly without the need for bacteria to make multiple copies of a particular stretch of DNA.

Pieter had sent his postdocs to learn the BAC technique at the California Institute of Technology (Caltech), and Pieter's lab was chosen, along with the lab at Caltech, to prepare the libraries for the Human Genome Project.

Within a week of the ad's appearance in *The Buffalo News,* sixty respondents had arranged to visit Roswell Park Cancer Institute. There, they would meet with a genetic counselor and give informed consent. Afterward, they would provide a 50 ml blood sample, which, from that moment on, would be identified only by a participant number and their gender. (In a twist befitting a genome project, the technicians who drew their blood were two sets of identical twins: one set Caucasian and the other African American.) From these sixty samples, human genome libraries were made at each site and then shipped to sequencing centers around the world who had been chosen to contribute to the Human Genome Project.

Although it was originally planned to use DNA sequences from ten randomly selected individuals, not everything went as planned. The Roswell Park libraries produced consistent and high-quality DNA, while the libraries from Caltech suffered from recurrent virus infections. For that reason, most of the human reference genome derives from just two libraries—meaning DNA from only two people—both made at

Roswell Park. Indeed, approximately 80 percent of the final sequence of the Human Genome Project comes from a library known as RPCI-11, containing the DNA of an African American, while the rest comes from RPCI-13, a library made from Pieter's own DNA.

You'll recall that in John West's genome, we couldn't find his Factor V Leiden clotting mutation. Confusingly, at that very same position in his genome, a different variant was reported as standing out: and it was one we quickly recognized as the healthy DNA sequence for that location. It occurred to us that perhaps the "reference" genome actually contained a mutation, a DNA letter that is associated with disease. After all, the reference came from real human beings. What if the human being who supplied the DNA for that section of the reference was actually affected by the Factor V Leiden variation? In fact, what if that stretch of DNA named for the Dutch city of its discovery actually came from the Dutch genome scientist whose DNA was one of the two successful libraries used in the Human Genome Project? It wasn't so unlikely a proposition, given that 3 percent of Dutch people carry that variant.

We checked the human reference and found our intuition was correct. The reference call for that position was, in fact, the Factor V Leiden mutation. So John's second, "normal" copy of the gene was being called a mutation. (He had one normal variant and one Factor V Leiden variant.)

This immediately raised the specter of making a serious error. What if a different patient had inherited two Factor V Leiden variants, one from each parent (again, not that unlikely for a variant present in up to one in twenty people in some populations)? This individual would have an even higher risk of blood clotting—possibly as high as eighty-fold the risk of the general population—but our fancy new genome sequencing test wouldn't detect it at all, since both mutant gene copies would have the same letter as the one in the human reference. This was breathtaking. And clearly, we needed to fix that.

More concerning to us, this would not be the only position in the genome where this was a problem. The reference was sure to harbor other risk variants for disease, just like the real people from which it was derived. We had identified the Factor V mix-up in John only because we went looking there on purpose. How would we find all the unknown others?

In the end, after much brainstorming, Rick Dewey, the new postdoc in our team, suggested an elegant idea. It is a pretty good bet, he reasoned, that for any given genetic disease, most of the population does not have it. Why not identify, then, every position in the reference genome where the reference wasn't the most common letter in the population and replace it instead with the most common variant? This could create, in essence, a disease-free reference. After all, if it's supposed to be a reference, wouldn't you want it, as with every other measurement in medicine, to represent the largest subsection of the "normal" population? It turned out a project known as the 1000 Genomes Project had started pooling samples from a few hundred individuals, sequencing those pools, and releasing the data. (We were still only talking about tens of people in the world who had ever had their genomes sequenced in depth, but this project provided information across three ancestry groups of the most common variations in those populations.) Using this data, over the next couple of weeks, Rick created three new human reference sequences, tuned to three major ethnicities, offering improvements in accuracy and speed for our pipeline, and minimizing the chance of missing serious, disease-causing variants. In the case of Factor V specifically, it meant we wouldn't be missing patients at eightyfold risk of pulmonary embolism.

Having now spent weeks on just one DNA position in John's genome, it was time to scour the rest. And not just in his genome, it was time to broaden our work to the genomes of the rest of the family.

. . .

We set to work on methods to leverage the "power of the family" beginning by calculating who inherited which piece of DNA from whom. We started by implementing a statistical approach called the *Hidden Markov Model* (HMM, Andrey Markov was a Russian mathematician from the turn of the twentieth century). This is a powerful approach that Carlos Bustamante's group had a lot of experience with. The HMM outputs a probability of a hidden or unknown "state" given several inputs that also include the previous state. It is a little complex, so bear with me while I try to take you through it. Imagine two friends who live in different cities. One friend (let's call him Buffalo Bill) has three ways of getting to work, and he chooses among them depending on the weather: walking to work, biking to work, and driving to work. Bill's friend Jill from Jacksonville magically gets to hear how Bill gets to work every day and even gets to hear what the weather was like in Buffalo over the last few days. Her task is to guess what the weather is like in Buffalo today. How exactly might she go about doing that? Well, let's say she finds out that Bill drove to work today, and let's say she estimates the probability it's raining today in Buffalo given it rained yesterday at 50 percent. She also guesses that it's 80 percent likely Bill takes the car if it's raining on any given day. Now, she can make an informed guess of the weather today in Buffalo (news flash: it's raining!).

Converting this to genome speak, if we start working our way along the genome sequences of the family and the task is to guess which child got which letter from which parent at each position, then the thing we know—Bill drove to work—is now represented by the known DNA sequence for each family member, while the weather today (the unknown) is who inherited what letter from which parent. It's a useful approach because which parent donated the series of letters just behind you on the road (the weather for the last few days) is actually an important determinant of which parent gave the next few (because parts of the genome are inherited together), just like for the weather.

So we made it. Now you understand one of the most useful mathematical models in all of science. What can you use this information for, though? One of the most useful things about knowing the inheritance state for any specific spot in the genome is that you can begin to do something called *phasing*—working out not just the two letters at any point in the genome that a person has but the *strings* of letters that are inherited together on the same chromosome from each parent. There are certain circumstances where it could be very important to know that. For example, if you find two potentially disruptive variants in the same gene in the same person, it is very important to know if they are on the same copy of the gene or on different copies. If both are on the same copy, then you still have one intact gene left and, it turns out, for many genes, that is plenty. But if both copies were hit by one disruptive variant each, well, now you have no working copies of that gene. Using this technique and a few other tricks, Rick was able to completely phase the genomes across the entire family.

As well as stretches of DNA inherited from each parent, every individual has a small number of brand-new variants. At that point in time, around 2011, no one really knew exactly how many. The answer to the question "How many?" is "Not many," because the process of copying DNA exhibits remarkable fidelity. It turns out that in making a human, a new variant arises approximately one in every one hundred million bases (the rate rises with increasing parental age) so that each of us carries about forty to eighty brand-new variants. This is part of what makes each of us genetically unique. Of course, the fact that the genome is very long (six billion letters) and only 2 percent of the genome is made up by the genes, means that there is only a one-in-fifty chance that one of those new variants hits a gene and a much lower chance again that the new variant hits an important region of an important gene. In the case of John's children, thankfully, none of the new variants fell in gene regions important for health and disease.

For all four West family members, Atul Butte's group had created estimates of risk for dozens of common but impactful diseases like high blood pressure and heart attack. These are sometimes called *polygenic* risk scores, because they are derived by combining the effects of many common gene variants together. Each low-impact variant, individually, may have only a small effect on the likelihood of disease, but together, they can paint a meaningful picture of risk. At the time, our knowledge of which genetic variants to include, and how to combine their effects into a final score, was quite preliminary. Still, by graphing each family member's risks on a background plot of the population risk distribution, we learned many things. One surprising reminder that came from these scores was that while, for some diseases, the children's risk scores fell around or in between the parents' risk scores, for others, they were significantly higher or lower. Just like every now and again, two shorter parents have a child that is taller than either parent. This was important from the perspective of judging genetic risk from asking about family history alone. Clearly, there was going to come a day when we were going to rely less on asking patients about their family medical history and more on measuring their actual genetic risk.

Despite our powerful computers, much of the work involved in clinical genome analysis involves human labor. Smart geneticists, genetic counselors, curators, and data wranglers bring together the evidence that a candidate gene is capable of causing disease and that a discovered variant actually causes disease. This process involves real people looking up papers and synthesizing what they see into an argument for and against each candidate variant. Each variant is then *prosecuted*, like in a court of law. Except instead of a courthouse, it's usually a low-ceilinged conference room, the witnesses called are scientific papers, and we are, ourselves, the judge and jury. Even here, though, computers can help.

We had already been thinking about how databases of genetic variants should be structured to allow individuals to specify how much they wanted to know. Would people who had their genomes sequenced want to know, for example, if they had genetic variants associated with diseases that had no effective treatment? How should we treat an intervention that could save a patient's life but involved invasive surgery, such as the implantation of a defibrillator or a bone marrow transplant? We started to catalog these factors in a standardized format, imagining virtual knobs or dials that could be turned up or down by individuals who had been sequenced so they could personalize what they wanted to know about their medical genome. The filters included things like age of onset of the disease, severity of the disease, availability of a treatment, invasiveness of treatment (for example, open heart surgery versus a daily pill). A critical aspect underlying such an approach was the confidence that a given variant actually *causes* a disease. At the time, there was nowhere to go to look up the "case law" contained in all those prosecutions of different variants by teams mentioned above. The challenge of cataloging those variants was taken on by a major NIH-funded consortium called ClinGen (Clinical Genomes) headed by an all-star group, including Heidi Rehm at Harvard, Sharon Plon at Baylor, Carlos Bustamante at Stanford, and Jonathan Berg and James Evans at the University of North Carolina.

Around six months after John West first called me, it came time to return the results to the family. Kelly Ormond, our genetic counselor, once again joined the scientists and physicians. We started by reestablishing what the family wanted to know. And not surprisingly for the Wests—John, Judy, Anne, and Paul—it was everything. Kelly discussed the fact that it wasn't usual to return results that relate to adult-onset diseases to kids under the age of eighteen. In addition, John and Judy felt they would like to hear about the kids' results before the kids did.

However, Anne and Paul had both spent time analyzing their own data and begged to differ; they wanted to know everything.

In the end, we presented a summary of our analysis to the family over the course of about two hours and followed up with lots of files and spreadsheets for them to peruse off-line. We started discussing what we had learned about human inheritance, and then we moved on to talk about specific medical findings in their genomes. Not surprisingly, we spent a lot of time talking about blood clotting.

Starting with the clotting-related variants we already knew John had, we had extended our search, once again, to all the genes and variants in the literature that were known to affect blood clotting for the other three members of the family (including Judy, John's wife). In addition, we looked at everyone's genetic ability to process medications and looked especially closely at warfarin, the anticoagulant drug that had failed to prevent John's second pulmonary embolism. The final family list contained five variants of relevance for blood clotting. We found that John had passed both of his risk variants to his daughter, Anne, but neither to his son. Surprisingly, we also found that Judy carried three variants that were known to predispose to blood clots, and she had passed two of them on to Anne (although neither she nor Anne had a history of blood clots). So Anne received a total of four variants predisposing to clotting, two of which came from her father, who had suffered blood clots, and two of which came from her mother, who hadn't.

What could the genomes tell us about their likely response to medications (pharmacogenomics)? Did Anne and Judy have a normal ability to metabolize warfarin in their livers? They did. We also found that Anne and her mother were ultrarapid metabolizers of another drug, called clopidogrel, that is sometimes used to thin the blood in patients with heart attacks or strokes. Since the drug is metabolized in the body to the active drug, ultrarapid metabolizers actually have more effective blood thinning and higher risk of bleeding. That meant that if Anne or Judy were given this drug in the future as a result of a clot, for example, in a heart artery, they would need to be particularly careful about bleeding.

Finally, we looked at John's genome to see if we could find a reason why, despite taking good therapy for blood thinning, he suffered a second blood clot. Was there a problem with the way he processed warfarin? Or did he just have a clotting system that was too strong for the warfarin doses he was taking? Our pharmacogenomics team examined the genes that were most relevant to John's metabolism of this drug and every other important class of drugs. They found he had genetically normal metabolism for warfarin, suggesting that his second pulmonary embolism was likely not caused by ineffective blood thinning but by his genetic predisposition to blood clotting.

When we published the results in 2012, John and his family were hailed as genome pioneers in *The Wall Street Journal.* Actually, it was hard to argue with that, particularly given their dual roles as patients and members of the scientific team. The West family were helping to usher in a new era, in which access to genome information was readily achievable (the Ferrari price on their genomes was a mere $1.50). It was not a question of whether you *could* learn this information but rather if you *wanted* to learn it. In fact, although she was still in high school, Anne would, over the course of the next few years, deliver several presentations to scientific audiences, where she would talk openly about the analysis and findings and how she believed they had already been relevant in her life. "Knowing that I have Factor V Leiden has helped me avoid medical treatments that could cause me to have blood clots," she said to me in the fall of 2019. "And knowing that I do not have variants in the genes for breast cancer has also been important." Now a law student at Indiana University with a special interest in medical privacy, she says she also appreciated learning about her family's genetic "quirks"—like where their freckles came from or why her dad hates cilantro. "Despite the staggering amount of information we learned from our analysis at Stanford, my family will need to keep our eyes open for new discoveries in the field of genetics that could have implications for us," she added. "Using personal genomics in medicine isn't a onetime thing, it's an ongoing journey."

# 7

# Starting Up, Reaching Out

*Those who travel outward seek completeness in things;*
*those who gaze inward find sufficiency in themselves.*
—LIEZI, *The Book of Master Lie*

*You're almost always better off*
*making your business better than your pitch better.*
—MARC ANDREESSEN, COFOUNDER OF ANDREESSEN HOROWITZ

Ever since our analysis of Steve Quake's genome in 2009, we had begun receiving calls from individuals interested in analyzing their own genomes—and not just John West: parents with kids affected by an elusive genetic disease, biotechnologists with early access to sequencing, and billionaires interested in access to the latest thing. With costs continuing to fall, it seemed clear that offering patients access to their genome would soon become much more common. The logistical route to eventually sequencing "everybody" was less clear, despite that common exhortation, but there already seemed to be a strong case for sequencing patients with rare or genetic disease. Genome sequencing was about to go mainstream.

At that point, our own sequencing efforts had been firmly rooted in academia, depending on graduate students and cardiology fellows, and frankly, we didn't have time to keep up with the demand we were facing. So in 2010, I started talking to Russ Altman about whether we should form a company to move clinical sequencing out of academia and into industry. He seemed interested. I also spoke to Mike Snyder and Atul Butte, who had already independently been talking about starting a company with this same purpose. Other than a website I built to sell my father-in-law's artisan metalwork (the now defunct funkymetals.co.uk), I didn't have any experience starting a company. But I knew someone who did: venture capitalist Michael Moritz.

Michael is a Silicon Valley legend: a Welsh boy from Cardiff and Manchester United superfan, who, before shepherding investments foundational to the Silicon Valley we know today, authored two books on the history of Apple, as well as one on leadership with Alex Ferguson, the longtime manager of Manchester United. Mike had attended Oxford and his college was, like mine, Christ Church. He was kind enough to take a meeting, and I went up to see him at his office on Sand Hill Road, abutting the Stanford Campus.

Sand Hill is the place in Silicon Valley where the venture capitalists live. From campus, it presents a moderately steep uphill climb by bike, and the trip "up the hill" is a pilgrimage taken by many Stanford faculty. You might have heard of some of the companies that resulted from such a trip: Google, Nvidia, Netflix, Tesla, Electronic Arts, Cisco, Dolby, Yahoo!, Instagram, LinkedIn, Snapchat, VMware, to name just a few. Michael is senior partner at Sequoia Capital, which is perched at the very top of the hill, and one of the most storied venture firms in the Valley. Over the past forty-eight years, Sequoia has invested in more than 250 companies that now have a combined value of $1.4 trillion, including Apple, Yahoo!, Google, Intel, Oracle, PayPal, Stripe, YouTube, and many more.

Well aware of this history, I arrived there on a warm summer day in June and waited, a little sweaty and rather nervously, in the plush lobby. The décor was sparse but tasteful and reeked of success. All around, I could see multiple conference rooms with expensive-looking furnishings and (at least compared to our weathered academic seminar rooms) preposterously large flat-screen TVs. Michael wandered over casually dressed in slacks and an open-necked shirt, as is the uniform of Silicon Valley, and, with a wide smile, held out his hand. We collected some chilled water from the refrigerator and headed to his office. At the time, open office spaces were all the rage, and so his office was in fact a desk. I was shocked to see that it was situated in the middle of several others, with colleagues just a few feet away. Here was a billionaire investor, included on the *Forbes* Midas list several times for being the most successful investor in Silicon Valley, sitting like a graduate student in the middle of an open-plan office.

"Well, I don't know if you have heard much about the revolution in genomics," I began, suddenly realizing what seemed revolutionary in my world might be less so to someone who cut checks to Steve Jobs (the cofounder of Apple), Jerry Yang (the cofounder of Yahoo!) and Larry Page (the cofounder of Google). He then listened attentively as I recounted, awkwardly and quietly at first, our various adventures in the genome space and what we envisioned a clinical genome sequencing company might look like. Illumina was already performing clinical-grade genome sequencing, he reminded me. Why, he asked, did this new company need to exist? He was warm and generous with an instinctive grasp of the topic, and I grew more animated as he put me at ease. I explained that we had built tools for the comprehensive medical interpretation of a whole genome, something we didn't think anyone else was doing yet (Illumina, for example, was more focused on sequencing rather than interpretation). I showed him the visualizations integrating medical and genomic data we had built for Steve's genome: "One day, your electronic medical record will look up your genome as easily as it looks up your medications!" I exclaimed, getting more excited now. He shifted focus and leaned back in his chair. "This does look interesting," he said finally. He was thinking. That seemed good. He agreed to introduce me to some colleagues in his own firm who could start to plan what a genomics company might look like.

. . .

Planning a company is exciting, if uncharted, territory for most academics, who live in a world that is not well connected to industry. Indeed, at some universities, industry is viewed as "the dark side"—a place where academics go in mild disapprobation, never to return. The higher salaries offered in industry lead to the commonly held notion of an academic selling out. This despite the fact that, in the United States, industry spending on basic research has actually outstripped that of the federal government since 2015. Partnership between academia and industry is steadily increasing. Yet despite this, the two worlds remain far apart.

Stanford stands as the antithesis to the two-worlds view of academia and industry. Over the years, the Stanford Office of Technology Licensing has become famous around the world as a model for the translation of intellectual property from the labs of Stanford faculty into industry. Since 1970, it has formalized the creation of more than 11,000 inventions and issued more than 3,600 licenses to commercialize them. All told, this has generated more than $1.7 billion in royalties. Some companies' technology was so intimately tied to Stanford, in fact, that the university was embedded in the company name: Sun Microsystems (named for Stanford University Network), for example, which birthed the Java programming language and was sold to Oracle for $7.4 billion in 2010. Sun's first computer was based on a workstation designed by Andy Bechtolsheim when he was a graduate student at Stanford.

In the biomedical domain, it was the Bayh-Dole Act, signed into law by Jimmy Carter on December 12, 1980, that enabled wide-scale technology transfer to industry. This act specifically allowed universities in the US, for the first time, to benefit financially from ("pursue ownership of") intellectual property created using federal research funds, such as those from the National Institutes of Health. Famously,

Bayh-Dole enabled Stanford to license the so-called gene splicing patents on restriction enzymes to Genentech in 1981, giving birth to the biotechnology revolution.

It was probably in my first week at Stanford that I noticed the difference this act created between the academic culture in the United States and the UK. Back in the UK, I knew one person with a patent, and he was mildly famous within our department for that. In my first week at Stanford, the word *patent* arose so many times I lost count. That may have skewed high because of my mentor's close collaboration with Agilent Technologies, one of the most patent-avid companies in biotechnology, but the difference in approach to protecting ideas for commercialization was profound.

Now as we moved toward launching a genomic diagnostics company ourselves, I was excited to explore further that culture. In relation to our new, still theoretical, nameless company, we started discussing the advances that could be licensed out of each of our groups, and started talking to more investors on Sand Hill Road.

One thing was immediately clear: although we hoped our genome team could be a strong scientific foundation for a company, we needed a business lead. "You will have no trouble convincing investors as Stanford faculty that your technology is good, but you have to convince them you can look after their money," was a refrain we had heard from many venture capitalists. Back at Stanford, we racked our brains, searched our contact lists, and asked a few knowledgeable people for ideas, but none of the names generated much enthusiasm. Then one day, it occurred to us that the answer was quite literally staring us in the face: John West. The CEO whose genome we had just worked on. Wasn't John a hugely successful engineer and businessman? Hadn't he played a pivotal role in the sale of foundational clinical-sequencing technology to the company now dominating the market? Hadn't we talked with him constantly over the last few months about the potential impact of the genome on clinical medicine? We decided that whoever saw John next would discreetly ask him if he would ever consider heading up such a venture.

As it turned out, a few weeks later, I was the next person to meet him at one of our regular meetings to discuss how our analysis of his family's genomes was progressing. So I prepared the pitch in my head as Rick Dewey laid out our latest insights into disease susceptibility in his family. When that meeting ended, I made my way toward John so I could ask to chat privately. Yet before I could open my mouth, he asked if he could speak privately to me for a moment. We stepped outside the conference room and did one of those "you first, no, you first" stand-up routines. I won, and he went first: "I was just wondering, you sometimes hear about academics doing this, have you ever thought of starting a company focused on genomics?"

We used to joke that he was the most deeply vetted CEO in history. We knew him right down to his DNA.

Once John had joined, the key "napkin" meeting for the company (named for the inspirational plans often scribbled on napkins over coffee) took place at the Mayfield Bakery & Café in Palo Alto, where they served delicious, dark-roast Italian coffee and the best almond croissants this side of anywhere. So my taste buds were tingling that chilly winter morning as I drew up at 7:00 a.m. We were about to enjoy precisely none of that, however, because we'd forgotten to check what time the café opened. We sat outside, cold and shivering on the empty patio, having failed in our first organizational task as founders to find an actual indoor location for our first meeting. But we had launched a venture that would engage us, obsess us, and pull us all into exciting new territory over the next few years. We were warmed by a creative fire as the ideas flowed and plans for the company emerged.

However, our company still needed a name. We tried a few out for size. One my brother had suggested was warmly received: Concinnity Medical (*concinnity* means "the skillful and harmonious arrangement or fitting together of the different parts of something"). That was until Mike Snyder tried unsuccessfully to pronounce it three times, failing a different way each time. To uproarious laughter, the name was buried, never to be heard of again. In the end, it was Mara Neal, a friend of Russ Altman, who came up with the name at a dinner at The Village Pub, a restaurant not far from Stanford. Russ was explaining the new venture, and his friend asked what the company was going to do. He replied that we were planning to use human genomes for personalized medicine. "Well," she said, "you should try a Latin word." She suggested the word *personalis,* from the phrase *cura personalis,* meaning "care for the whole person," a refrain commonly heard in the Catholic church tradition. Russ promised that if we used the name, he would gift her one share in the company (which she claimed in December 2019, six months following the initial public offering).

In the end, there was a large portfolio of intellectual property transferred from Stanford to Personalis, a feat that took several months of negotiation between John and the licensing office. During that time, the founding group had meetings most Sunday afternoons, where we talked and dreamed and discussed how the various pitches to venture capitalists had gone that week. When it finally became clear we had secured an investment team, led by Jonathan MacQuitty at Abingworth (the same firm that had funded Solexa) and including Sue Siegel, a renowned biotechnology investor, who was at the time with Mohr Davidow Ventures, it was time for celebration. The bookend to that cold Mayfield Bakery & Café morning was a sunny Sunday afternoon in the conference room for the biomedical informatics graduate school program, where we toasted our venture with beers chilled in one of those standard-issue, rubber laboratory ice buckets that Mike Snyder had swiped from his lab. If you look carefully in the doorframe, you can still see the marks from our celebration—no one had a bottle opener, but some undergraduate skills you never lose.

It was a few months later, and Mike Snyder was on a mission to measure everything about himself, all the time, using every technology possible. And I mean everything. Starting in 2010, shortly after he started at Stanford, Mike would show up to meetings sporting multiple different wearable devices. You would meet him, and there might be one smartwatch on one wrist and a different one on the other. Sometimes, he would wear an armband device the size of a pack of cards that detected airborne toxins in his environment. At one meeting, he showed up with a front-facing camera that took time-lapse pictures of everyone in the room. It freaked everyone out, so he stopped that soon after. Lloyd Minor, the dean of Stanford's School of Medicine, refers to him as "the most studied organism in history."

Naturally, one of Mike's first acts was to sequence his own genome. He wasn't content to stop there, of course. He wanted to monitor which genes were most active in the white blood cells of his immune system (recurrently). He wanted to measure the proteins circulating in his blood and the breakdown products of energy production in his urine. He also studied his poop quite intently. His microbiome (the bacteria found on our skin and in our gut) was scrutinized deeply. He used to get quite excited when he experienced a biological perturbation, one certain to provoke a noticeable change in his measurements. I remember once he called me up, excited.

"Euan, great news!" he exclaimed.

"New grant? New paper? New project?" I wondered out loud.

"I've got a cold!" he replied, barely containing his excitement. "Can you come and take my blood?"

Genomics. Proteomics. Metabolomics. It was a one-man multi-omics show he was both directing and starring in.

My lab's role in Project Mike was clinical whole-genome analysis and interpretation, so cardiology fellow Rick Dewey fired up our pipeline again—the same one used to analyze the West family. In Mike's genome, we found a disease-causing variant in a gene called *TERT* that exerts its effect via shortening the telomeres (the protective ends of our chromosomes, the size of which can indicate cellular aging). In response to this finding, Mike, of course, actually measured the length of his telomeres and found the number of white blood cells in which his telomeres were shortened was at the upper end of the normal range. This meant his bone marrow might prematurely fail in its job of making red blood cells. So while Mike was more focused on his risk for diabetes (he first started showing signs of diabetes during the period of time he was studying himself), we were more concerned about his genetic risk for aplastic anemia—a rare disease in which the bone marrow stops making red blood cells. We recommended regular follow-up via blood cell counts as an early-warning system for anemia.

In 2012, Mike published his findings in the journal *Cell* as one of the first multi-omics studies in the literature. The fact that the several billion data points summarized in the paper were all measured in the senior author led to some unusual coverage. *Nature News* even coined a new word in its headline: THE RISE OF THE NARCISSOME.

Yet Mike wasn't only interested in his own biology. His vision was to bring genome analysis to everyone, including every patient at Stanford's hospitals. He had even offered free genome sequencing for any genetics faculty who were interested (I was already working on sequencing my own genome, along with that of my parents, but many other faculty signed up to do it this way). Meanwhile, he had also started a pilot project to sequence real patients in the clinic with my postdoc mentor, Thomas Quertermous. TQ was the man who initially brought me to California ("Okay, we'll take your Scot!" was the note he sent to Vic Froelicher, five words for which I am, to him, eternally grateful). Mike and TQ had recruited patients from Stanford's primary care clinics into a research study focused on returning whole genome sequencing results. A dozen participants had now been enrolled. Time was passing, and the sequence data was coming available. We needed a plan to interpret and return the results.

Since we had worked on Steve Quake's genome, we had been envisioning bringing genome sequencing to Stanford clinics more broadly. This study was our first real attempt to show how this technology could be used in a real-world general medical clinic. Running the pipeline to identify gene variants was routine by this point, but to fully interpret a genome for medical relevance still took around fifty hours of human labor. With ten or more genomes coming, sequenced both using Illumina technology and that from a relatively new rival called Complete Genomics, patients were waiting to hear about their results, and it was clear we needed help. What we needed was a genetic counselor with training and experience in analyzing and prioritizing genetic variants—someone with exposure to genome sequencing, someone who was ready to start straightaway and who wouldn't be fazed by an ambitious and minimally articulated plan. What were the chances of finding one of those? Genetic counselors were, by now, hot property, with demand far outstripping supply, and those trained in a program with exposure to genomes and exomes, as they were at Stanford, even more in demand. I called the lead genetic counselor in our Center for Inherited Cardiovascular Disease, Colleen Caleshu, for advice. I couldn't have been more surprised when she told me, "You know, I have just the person."

Megan Grove was, at the time, a new graduate of the Stanford genetic counseling program. A former Stanford undergraduate sailing team captain and a perpetual radiator of positive energy, she had also memorized more fist-bump combinations than potentially any human in history. She appeared perfect for the job: an unusual combination of talent, energy, optimism, and a serious nose for adventure. So with Mike producing the sequence data in his lab, TQ communicating with the patients, Rick running the pipeline, and Megan ready with interpretation, our next genome adventure was beginning.

Things had changed a bit since most patients had signed on to the project, so our first task was to contact them to update their original consent. Fortunately, this would give Megan an opportunity to deliver some genetic counseling, setting expectations before we delivered results. Many patients of the Stanford primary care clinic we had chosen for the study were local to Stanford, but many also had the means to live and travel all around the world. And so, a complex phone tree exercise began to track down each participant, update their informed consent, and deliver genetic counseling. We also let them know that data was coming.

The sorts of insights we were hoping to glean from that data and return to patients were similar to those we had derived from earlier genomes, but the process was more streamlined now. We had a vastly improved rare variant pipeline that Rick was beginning to link together in a "wrapper" so it could be run in one step. This allowed a filtered set of prioritized variants to be delivered to Megan, our one-woman curation and counseling team. Meanwhile, we updated our polygenic risk scores for heart attack and diabetes. We also integrated those genetic scores with the traditional risk factors, such as high blood pressure, smoking, and cholesterol. The pharmacogenomics team swung back into action. What was new now was a formal genome report. This was a succinct summary of a full analysis of a patient's genome with respect to all the genetic risks we were able to understand at that moment in time.

As we began to try to tackle this many genomes at once, it was immediately evident there were several ways in which genome sequencing had to advance to be ready for prime time in patients. The first was simply in coverage—the number of independent times the sequencing process covered each position in the genome, checking and double-checking to see what DNA letter was located there. We had learned early that the number of times we read each position was a critical component of whether a confident A, T, G, or C could be called at that location. We eventually coalesced around a standard of at least twenty high-quality bases properly mapped to the location of interest.

But in the fast-moving world of genomic technology, that standard had not always been achievable to date. In Steve's genome, we read each position an average of twenty times, while in the West family, it was around thirty. Under these conditions, our question was: For any given patient, at how many positions in each key gene could a call confidently be made? We decided to check the fifty-six genes that the American College of Medical Genetics had indicated were among the most important genes (because they could provide actionable insights into conditions such as inherited cancer or sudden cardiac death).

The results were sobering. We found that every single one of those medically important genes contained significant areas that could not be confidently called. This meant that a disease-causing variant might lurk in one of those areas, but we would be unable to find it. In most cases, it was 2–6 percent that could not be confidently identified—but for some genes, it was far in excess of 10 percent. This seemed like a big problem. If you are sequencing patients and miss 5 percent of a gene, you may miss the opportunity to make, or rule out, a diagnosis: something that could be relevant quite literally for life or death.

In contrast, other findings from our genome pilot provided significant reason for optimism. There had been a lot of worry among genetics professionals about the consequences of sequencing "healthy" individuals, not least that the downstream medical costs would be unaffordable. So as part of our study, we tracked those costs by asking the primary care physicians for each patient what tests they wanted to perform to follow up on the genetic findings we presented to them. We tallied up the charges for these procedures and found that the average total was not in the tens or hundreds of thousands of dollars, as some had predicted, but rather was around $700. That was relatively modest, even before considering the fact that genome sequencing might *save money* on other expensive medical treatments.

One of our participants was a good example. She harbored a disease-causing variant in *BRCA1,* a well-known gene that when mutated can cause breast and ovarian cancer. The implications were clear: her risk of future breast or ovarian cancer was high. Indeed, by some estimates, this variant increased her lifetime risk of breast cancer to over 50 percent (the average for someone like her otherwise was 10 percent). Some patients with these kinds of variants choose to have their breasts or ovaries surgically removed to reduce the chances of cancer to almost zero, a topic the actress Angelina Jolie highlighted in May 2013 when she discussed, in an op-ed in *The New York Times,* her personal decision to undergo mastectomy, because of her own *BRCA1* mutation.

Megan and I picked up the phone to break the news to our patient. It was a hard call to make, but so much harder to receive. We huddled around the landline phone in my office and, over a couple of hours, took our patient and her husband through the finding and what it meant. We started by explaining again how the sequencing had been performed and what our process had been for analysis. We emphasized that this was a research finding at this point and would need to be confirmed with an approved clinical laboratory test. We then discussed how we had found the variant in *BRCA1* and went into some detail over the quality-control measures we had taken to reassure ourselves that this was real. We paused, aware that neither our patient nor her husband were going to remember much beyond the headline news we had delivered. What were their questions? We explained how the *BRCA1* gene is involved in DNA repair and how, when it is faulty, dividing cells are more prone to accumulate cancer-causing mutations. We talked about how this translated to increased risk of breast and ovarian cancer. We described the surgery that some patients choose to undergo to mitigate this risk. Finally, and most importantly, we described how our breast cancer team was ready to talk and that we could organize an immediate appointment. They gratefully took us up on that offer.

It was clear to us, from that moment, that we were entering a world where access to this kind of risk information, hiding in plain sight in our genomes, would now be possible, mundane even. As with any new technology, the greatest challenges would be in the ethics and economics of its use.

The final paper, published in the *Journal of the American Medical Association* in the spring of 2014, focused on both the technical challenges we faced in bringing genomics up to clinical grade, but also on the amazing opportunity presented by ready access to that kind of data. We laid out reassuring findings from our financial analysis, that the genome test would not add a hefty burden to the cost of medical care. Furthermore, given the country's limited supply of geneticists and genetic counselors, the study also yielded encouraging news about the feasibility of returning results via a general practitioner: the GPs welcomed the new information and the opportunity to address patient risks before disease took hold. Finally, we showed that potentially life-changing findings would emerge for a small number (fewer than one in ten individuals) such as our patient with the *BRCA1* mutation.

Just before publication, I spoke to Nancy Shute at National Public Radio, and I reached for an analogy to capture our sense of the promise and opportunity of the genome. A few short years after the birth of genomic medicine, we had gone from having only a handful of genomes sequenced in the entire world, with very little framework for their analysis, to the point where we could now think of offering sequencing to anyone with a rare, genetic, or unexplained disease. "We were all witness to the birth of this idea," I said, referring to the recency of the advances, "and now we feel like we have an unruly teenager on our hands."

"It's going to take some tough love," I concluded, laughing, with only a hint of irony.

Part II

# DISEASE DETECTIVES

# 8

# Undiagnosed

*You see, but you do not observe.*

—SHERLOCK HOLMES, "A Scandal in Bohemia"

ARTHUR CONAN DOYLE

*Dr. Eric Foreman: I think your argument is specious.*

*Dr. Gregory House: I think your tie is ugly.*

—*House,* SEASON 1: "OCCAM'S RAZOR"

Like many humans, I find my fellow *Homo sapiens* endlessly fascinating. I love watching their movements, their interactions, the way they act when they think they're not being watched. I sometimes pretend I'm the BBC's wildlife correspondent Sir David Attenborough and provide, sotto voce, a gravelly commentary on the behavior of these creatures in their native habitats. I'm actually secretly convinced everyone does this.

People-watching is especially fascinating when traveling to different cities or countries. I like the universality of human habits. I try to guess the relationship between random strangers, why they came to this place at this time, and what they do when they're not here. Is that an accountant checking a little too carefully every item on the restaurant bill? Or maybe a photographer or painter sizing up their next landscape as their gaze lingers in the distance? My mother-in-law is particularly good at this game. She complements her uncanny knack for visual observation with eavesdropping that is occasionally even subtle. Sometimes, we're out at dinner and her eyes glaze over in the middle of one of my stories, and I realize she has become absorbed in the conversation between the older lady with the blue rinse and the younger gentleman with the slicked-back hair at the table behind me. I switch my auditory attention to try to keep up, but she operates this game on a much higher level.

As a physician, such observation takes on an extra dimension. The beauty and torture of a traditional medical education is that it is impossible not to see diagnostic clues in the humans all around you: the curved nail indicating iron deficiency anemia, the flushed cheeks suggestive of lupus, the spidery blood vessels or yellowed eyes indicating liver disease. Many genetic conditions, in particular, produce characteristic changes in physical appearance: the slope of a nose, the spacing of the eyes, the height of the ears. Does that tall, skinny person have Marfan syndrome, a genetic condition of connective tissue? Or that person I was just introduced to who took a little too long to release my hand from their handshake. Were they exerting some primal authority over me, or could that be the early stages of myotonic dystrophy, a brain-muscle disease? Such is the relentless inner voice of the diagnostician.

Combining active sensory interrogation with a deep understanding of people is where the fun really begins. In clinic, this means reading the room. Exquisitely manicured nails and detailed care with appearance down to the last accessory? This could be personal style. Or just in case, dig a little deeper with your questions in the event this is actually a stoically prepared veneer of artificial wellness. Irritation and constant checking of the clock? Maybe recheck that blood pressure, think about the thyroid gland, and ask about work-life balance (for it will rarely be volunteered unless you ask and leave space for the answer). Familial interactions not entirely smooth? Perhaps there's a family dynamic here, barely hidden, that will frustrate the patient's attempts to stop smoking. You should turn your attention to the patient's spouse, without whose support the patient will never stop.

There was a great show on the BBC that I used to watch during my lunch break while writing my Ph.D. thesis. It was called *Through the Keyhole*. The intrepid presenter would lead us through the rather too-tidy home of some random celebrity, pointing out furnishings here and there, and, after twelve minutes, urge us to guess whose house it was. The host's catchphrase was the highlight: "The clues are there!" Except he pronounced it with an affected, mid-Atlantic accent (half-way between American and English), further drawn out for effect: "The clooooozz. . . . arrrr . . . thay-rrrr!" I repeat this catchphrase in my head from time to time when I am working on a problem (it often makes me laugh out loud). The advice is sound, you see; the clues really are there.

Arthur Ignatius Conan Doyle was himself a physician famously fascinated by the power of observation. Born in 1859, he was educated in medicine at the University of Edinburgh in Scotland. After graduating in 1881, however, he struggled to find his place in the profession. He worked as a ship's doctor for a while, then opened his own general practice in Southsea, England, before finally specializing in ophthalmology and starting a practice in London. Despite his training, few patients came knocking at his door—which turned out to be a good thing for the world, as it left him time to turn to his other passion: writing. Over his lifetime, he wrote many fiction and nonfiction books, but he became best known for a series of fifty-six short stories and four novels that featured crimes solved by a "consulting detective." The first was published in 1891. Aided and fictionally chronicled by an ex–army doctor, the consulting detective went by the rather curious name of Sherlock Holmes.

In developing the character of Holmes, Conan Doyle leaned heavily on the influence of one of his medical mentors, Dr. Joseph Bell, a surgeon at Edinburgh Royal Infirmary, whose famous patients included

Queen Victoria. Bell was widely renowned for his skills of observation, often randomly picking out strangers and guessing their profession and recent activities, to the amazement of those nearby. Of Bell, Conan Doyle said, "His strong point was diagnosis, not only of disease, but of occupation and character."

Given Conan Doyle's background, it is perhaps not surprising that the Holmes stories feature medicine prominently in various forms. As early as the first scene of the first book, Watson meets Holmes in a medical laboratory where Holmes is trying to improve the ability to detect blood at a crime scene by developing a test for hemoglobin. In fact, over the course of the complete works of Sherlock Holmes, there are references to sixty-eight diseases, thirty-eight doctors, twenty-two drugs, twelve medical specialties, and three medical journals.

Holmes's process requires observation, knowledge of the world, and an understanding of humanity. Many parallels can be drawn between it and how we approach medical mysteries. Holmes often makes the point that "seeing" (passive) is not the same as "observing" (active). In the story "Silver Blaze," Holmes picks out a half-burned matchbox from the mud minutes after the same area was searched by the police inspector, who remarks, "I cannot think how I came to overlook it." Holmes replies, "I only saw it because I was looking for it." In the same way, we teach our medical students to actively observe: if they feel a slow rising pulse, they must go looking for the murmur of a narrowed aortic valve.

Active observation extends to the absence of expected findings, as well. In "Silver Blaze," Holmes refers to the "curious incident of the dog in the night"—the curious incident being that the dog did nothing in the night. The fact that an intruder came by the stable and the dog did *not* bark was the key observation. The inference: that the perpetrator was someone well known to the dog. But you only detect the pertinent negative if you go looking for it. If I feel the distinct rise and fall of a "water-hammer" pulse on the patient's wrist, I go looking for the murmur of a leaking aortic valve. If I then do not detect it, as a diagnostician, I must investigate this "curious incident" further.

Holmes uses all his senses. In medicine, too, we have historically encouraged multisensory perception in our students. We no longer dip our fingers in urine and taste for glucose to diagnose diabetes, but we try to go beyond hearing and vision. Smell, for example, can be a particularly potent diagnostic aid. The ancient Greeks and Chinese used scent to detect infections such as tuberculosis. Today, every medical student is taught to distinguish the specific scent of ketones on the breath of a patient with diabetic ketoacidosis. No doctor or nurse can forget the smell of the stool of a patient whose stomach ulcer has leaked blood into the gastrointestinal tract. (The rest of the ward will usually not forget that either.) The power of the nose was not lost, of course, on Conan Doyle. In *A Study in Scarlet,* Holmes sniffs the dead body as part of his examination. In *The Hound of the Baskervilles,* he discusses holding a piece of writing paper within a few inches of his face and being conscious "of a faint smell of the scent known as white jessamine." This suggested to Holmes the presence of a lady, which dictated his whole strategy in the case from that moment.

Modern culture continues to highlight the medical allusions within the Holmes canon. Indeed, the creator of the Fox TV series *House* acknowledges a debt to Conan Doyle's famous consulting detective. That show ran eight seasons, from 2004 to 2012, and in 2008, it was the most-watched TV show in the world—a world, it seems, that is fascinated by a good mystery, medical or criminal, or preferably both. Several Easter eggs are hidden in the series for those paying particular attention to the parallels. At one point, for example, it is revealed that Gregory House, M.D., the main character, lives in apartment 221B and his driver's license later confirms that he lives on Baker Street, the same address as Sherlock Holmes. Both fictional characters have problems with substance abuse, and both rely on one close confidant, a doctor in both cases: one called Watson, one called Wilson. The credits for the episode where House is shot reveal that the name of the shooter is Moriarty (Holmes's nemesis).

The premise of *House* was that this renegade diagnostic unit in an academic hospital was willing to take on the hardest cases, willing to use any method possible to get to a diagnosis, even if that meant fighting rules, regulation, biology, and often the conservative medical establishment to save lives and find cures. There are a lot of similarities between this mission and the mission of the real Undiagnosed Diseases Network, brainchild of Bill Gahl at the National Institutes of Health.

William "Bill" Gahl, M.D., is about as close as you can get to a real-life Gregory House, M.D. Yet in contrast with the impatient, irascible, moody House, Bill is dependably kind, serious about serious matters, and frequently sardonic. Lightning quick with a wisecrack, he has a schoolboy's instinct for humor and appears to exist in a state of permanent readiness to produce one of his groaningly punning jokes.

Bill was born in the city of Waukesha, fifteen miles west of Milwaukee, in Wisconsin. His father was an English teacher at a trade school in Milwaukee, while his mother was a typing and shorthand teacher. He attended an all-boys Jesuit Catholic high school, where he studied four years of Latin, two years of Greek, and also learned, in his words, "to write and speak and things like that." He went on to MIT, where he realized almost immediately that he wanted to study disorders of the body's metabolism. By second year, he narrowed this desire to genetic conditions ("inborn errors of metabolism"). As part of his degree, he was allowed to take a genetics course at Harvard and, in his spare time, he worked at MIT in the lab of John Stanbury, the grandfather of the field, and the original single author of the classic textbook *The Metabolic Basis of Inherited Disease.* And so, before he had taken one step across the threshold of medical school, the die was cast. Bill entered medical school at the University of Wisconsin–Madison in 1972 and enrolled in a Ph.D. program at the same time. He followed that with four years of pediatrics, then headed to the NIH to specialize further. There, he did lab research on a rare metabolic disease while working on the board exams for clinical and biochemical genetics. Since there were fewer mandatory requirements for these certificates at that time, he supplemented his curriculum by looking for interesting patients in the NIH Clinical Center.

That experience, roaming the corridors of the NIH Clinical Center by day and studying metabolic disease in the lab at night, inspired Bill to think bigger. He wanted to tackle those stubborn cases that remained undiagnosed. In fact, in 1989, he proposed to the National Institute of Child Health and Human Development a plan whereby he would dedicate a significant portion of his time to the study of "new" diseases. "Bill, that's a terrible idea, and it will ruin your career," one mentor told him at the time, he recounted to me, with one of his deep belly laughs. He returned to his day job, researching and seeing kids with rare metabolic diseases.

Almost twenty years would pass before he had a chance to put his plan into action. In 2007, the head of the Office of Rare Diseases Research at the NIH called him to say they were being contacted regularly by patients with undiagnosed disease. The rare disease office was supposed to be the place you went when you had a *known* rare disease, he pointed out. These patients with no diagnosis were falling through the cracks. They had some modest funds available, and Bill was asked if he would use those funds to try to help these patients. He was intrigued. This was something like what he had in mind all those years ago. By coincidence, around the same time, there was renewed interest from the NIH Clinical Center in expanding their "fascinoma" clinics—clinics where people with highly unusual and unexplained "fascinating" findings would go. These two ideas seemed to dovetail well, and all of a sudden, after twenty years, it looked like his original idea might come together. The NIH's own Dr. House was born.

These efforts might have remained relatively modest, however, were it not for the vision of the NIH director at the time. Elias Zerhouni was born in Algeria but spent most of his career at Johns Hopkins University, where he rose through the ranks, becoming chair of the Department of Radiology then vice dean for the School of Medicine before being appointed as the director of the NIH by President George W. Bush. At the time, Bill did not know that the NIH director had even heard of the new undiagnosed diseases effort, though he started to get the idea when asked by the director's office to help draft a press release to announce it. It finally dawned on him what he might have gotten himself into when he attended the announcement on May 18, 2008, in the medical boardroom of the NIH Clinical Center and found that twenty-five members of the press as well as representatives from ninety patient advocacy groups had dialed in to listen to the teleconference. This national attention for the new Undiagnosed Diseases Program led to features in *Newsweek,* on the NBC *Nightly News,* on the Discovery Channel, and many others. Zerhouni's vision was clearly stated in the press release: "A small number of patients suffer from symptoms that do not correspond to known conditions, making their care and treatment extraordinarily difficult. However, the history of biomedical research has taught us that careful study of baffling cases can provide new insights into the mechanisms of disease—both rare and common. The goal of NIH's Undiagnosed Diseases Program is two-pronged: to improve disease management for individual patients and to advance medical knowledge in general."

This hugely ambitious program, announced with some fanfare, was at that point made up only of Bill, two nurses, and a scheduler. Yet the national press had just alerted the whole country to the fact there was now a place for all those undiagnosed patients who had nowhere left to turn. "They just started sending the charts," Bill recalls. "I'm sitting at this desk and thinking, *What the hell did I do here?* I'm a pediatrician, and I have a six-inch-high chart [from an adult]. I was actually a little bit afraid." Fortunately, Bill had been at the NIH Clinical Center for many years, and so he started calling his friends and colleagues, appealing to their generosity and curiosity in equal measure. Over the next few months, he put the train on the tracks. What wasn't yet clear was where that train might be heading.

Despite the fanfare, funding was modest and the whole program relied on Bill's working long hours into the evenings, poring over charts and then imploring colleagues to help with difficult cases. Bill went back to Zerhouni to say he thought relying on the charity of his colleagues was unsustainable. Thankfully, the NIH director found Bill persuasive and, feeling the pressure of the positive press reaction to the program, offered to put in $1 million from his director's fund if the heads of other NIH programs would come together to match it. With almost $2 million, the Undiagnosed Diseases Program was, for the first time, on sound fiscal footing. Even better, although Zerhouni left the NIH shortly thereafter, before leaving, he ensured that an enhanced budget was in place for the next three years. It allowed Bill to bring on two pediatricians who were also clinical and biochemical geneticists, Cynthia Tifft and David Adams (who is also a bioinformatician), and metabolic disease nurse specialist Lynne Wolfe. But that was just the beginning. Soon, the program would expand nationwide.

I first heard about plans for a multisite Undiagnosed Diseases Network in 2013 from Bill's colleague David Adams at a genetics meeting in D.C., where the two of us chatted at the bar. The idea, he explained, was to involve academic medical centers. A grant from the NIH would fund clinical work to solve a rare undiagnosed disease, irrespective of a patient's ability to pay, then also fund the follow-up scientific research needed to understand and treat it. This new network had recently been approved for funding after Bill Gahl presented the idea to Francis Collins, the current NIH director. This was amazing. I had never heard of an NIH program like it. Indeed, it seemed to me that solving such "hard cases" was the reason many of us went into medicine. It sounded very much like something we at Stanford would want to be part of.

The application process for any NIH grant is fairly arduous, and this was no exception. There are informational websites to read, FAQs to dissect, and webinars to attend. Then you assemble your team, detail a budget, request support letters, write the proposal, rewrite the proposal, chase everyone down for their overdue contributions, and hope it all comes together in time. It's a chaotic and stressful dash to the finish.

The most essential task was assembling the right team. I knew from Bill Gahl's prior work at the NIH program that a lot of undiagnosed patients are children with brain diseases. I therefore called Paul Fisher, our charismatic chief of pediatric neurology, enlisting him to the cause. In addition, Jonathan Bernstein, a brilliant clinician and chief of Stanford's Division of Medical Genetics, had spent years in the trenches doing exactly this kind of work. With Matt Wheeler now on the faculty after finishing his cardiology training, we had our leadership team together. In our proposal, we tried to highlight all the elements we thought Stanford could bring to a national program: from our prior work on genome sequencing to new technology and innovations like telepresence terminals to bring distantly located experts "into the room" with rare disease patients. I'm pretty sure we even put a photo of a robot in the application. Whether it was the robot that sealed the deal or not, we were thrilled to be included in the initial group of centers. These included a network coordinating center at Harvard, seven clinical sites (Stanford, Duke, Harvard, Vanderbilt, UCLA, Baylor, and the NIH program), and two sequencing centers. Later, we would add a biorepository, two model organism screening centers, and, in 2019, five more clinical sites.

The NIH leadership was generous enough to ask me to cochair this new network with Bill because of our experience with genome interpretation. And so it was that, rather jet-lagged, at the equivalent of

4:30 a.m. California time, I found myself in a conference room at the appropriately named Helix Hotel in Washington, D.C., for the initial steering committee meeting with forty people, most of whom I knew only by reputation. After some quick handshakes and introductions, we sat down. Bill turned to me, "Shall we get started?"

The doors to the Undiagnosed Diseases Network were officially flung open on September 16, 2015—and the years since have been a scientific and intellectual adventure. In the chapters that follow, you will meet some of the patients we have cared for and see how access to their genomes has altered their lives. Working with these families has been an exercise in what I most love about medicine: tapping our inner Sherlock Holmes. It has also afforded us an intimate view of what patients with undiagnosed disease go through: the unique isolation and torment of suffering from a disease with no name. It is an experience aptly known as a *medical odyssey.*

The epic poems of the Greek bard Homer are not called *epic* for nothing. In *The Odyssey,* the hero Odysseus spends twenty years escaping man-eating one-eyed giants, hypnotic siren songs of beautiful women, monsters with dog heads replete with rows of shark's teeth, and multiple shipwrecks. He finally arrives home to find that multiple suitors have taken up residence in his house, hopeful for proof of his long-rumored demise so they can compete formally for the affections of his wife. After what must have been a deep sigh, he concocts a cunning plan to host an archery contest among the suitors, which he enters and wins, after which he kills all the suitors with what is left of his arrows.

Given what he went through to make it back to his beloved, it seems fitting that Odysseus's travels entered the common lexicon. The *Oxford English Dictionary* defines *odyssey* as "a long and eventful or adventurous journey or experience." And given how long these families toil before finding a diagnosis, if they ever do, it is no wonder that their journeys are compared to that of Odysseus. Spending years going from one doctor to another, riding a roller coaster of emotion, accumulating mountains of medical bills while expending tens or even hundreds of thousands of dollars, even more in emotional reserve, clearly qualifies as a long and eventful journey.

I cannot begin to comprehend the emotions these families experience, when the diseases involve devastating neurological symptoms, often in young kids, and when the uncertainty goes on for years. Many liken the state of being undiagnosed to being stranded on an island. Our role is to help build a bridge off the undiagnosed island, a way out of the self-doubt and isolation.

That is the hope we give to our patients when we make a diagnosis.

9

# The Luck of the Irish

τέτλαθι δή, κραδίη: καὶ κύντερον ἄλλο ποτ' ἔτλης.
*Be patient, my heart: for you have endured things worse than this before.*
—HOMER, *The Odyssey*

*Those who explore an unknown world are travelers without a map;*
*the map is the result of the exploration.*
*The position of their destination is not known to them,*
*and the direct path that leads to it is not yet made.*
—HIDEKI YUKAWA, PHYSICIST

Matt Might had a blog. He had learned how to survive a Ph.D. program the hard way, and he had a talent for communication. So he blogged about computer science, graduate school, and much more. He had a lot of followers. In late 2007, he and his wife, Cristina, had also recently given birth to a baby boy named Bertrand. The new parents, however, almost immediately began to notice some unusual things. "At six months, it was clear something wasn't right," Matt remembered as he sat in my office during a visit from Alabama to the West Coast. Matt is sandy haired, with a light beard and blue-green eyes that sparkled as he warmed to a story he's told thousands of times. It was early morning, but as usual, Matt was energized and dynamic. He maintains a words-per-minute count higher than most people, and his brain is clearly three thoughts ahead of his voice. Bertrand, he continued, was developmentally delayed and appeared to be having seizures. "And later on, we noticed that he couldn't cry. Actually, he could cry, but he wouldn't make tears."

Matt and Cristina went to pediatricians, to developmental specialists, to pediatric neurologists. Each would come up with an idea for an underlying cause, most of them fatal. "It was just this one diagnosis after another," Matt recalled, noting that after a series of tests each diagnosis would, in turn, be rejected. This roller coaster continued for two years. "It was just terrible, because it was, like, death sentence after death sentence after death sentence, and then he would keep getting a reprieve."

Matt was no stranger to overcoming complex challenges, but his had generally been of the computer coding kind. As a kid, at twelve years old, he had been a hacker, and that passion stayed with him through high school and on to college at Georgia Tech where he completed a bachelor's, a master's, and a doctorate in rapid succession. After graduation, he started two tech companies, and while the start-up life had advantages, steady health insurance was not one of them. It was becoming clear that Bertrand would need specialized medical help, so Matt needed a more stable job. He joined the University of Utah's computer science department with a research program focused on cybersecurity. He used government supercomputers to do things like simulate clean coal power plants. Probably better than hacking them.

Despite his proximity to supercomputing, he had not, however, cracked the code of Bertrand's medical problem. After almost two years, neither had anyone else. After many twists and turns, he and Cristina ended up at Duke University Medical Center. There, in April 2009, they met Vandana Shashi, Kelly Schoch, and David Goldstein, a geneticist, a genetic counselor, and a computer scientist, respectively, who had teamed up on a pioneering study in which they planned to do exome sequencing (sequencing the 2 percent of the genome made up by genes) to solve difficult medical cases. Matt and Cristina didn't hesitate: "Sign us up."

The following two years were hard while they waited for results. Bertrand needed a lot of care—wheelchair bound and lacking the ability to communicate, he would often grind his teeth or stare into space, making doctors believe he was having seizures. He would be hospitalized every few months, which caused a lot of stress, as both parents had to find a way to fit this unanticipated full-time role into their lives. Then, in 2012, they received a call from the Duke team. "We think we know what this is," Vandana told them. Although they had several possible gene variants on their short list, the team thought it most likely Bertrand's condition was caused by complete disruption of a gene called *NGLY1* ("en-gly-1"). If that were true, she added, he would be the first person alive to be diagnosed with that condition. Bertrand had inherited one disrupted gene from each parent. So while his healthy parents each had one disrupted copy plus another working copy to protect them, he had no working copies. The gene was thought to be involved in the cell's "trash collection" machinery, and if that process were somehow halted, the "trash" would be expected to accumulate in his cells—causing disease. The Duke team followed up by looking at a biopsy of Bertrand's liver tissue under a microscope. There, they did indeed find an unidentified substance accumulating. Although it was hard to tie this observation directly to the gene, it seemed to make sense. But then came the kicker: "You know, we think this explains it, but in the absence of a second patient, it's really hard to know for sure."

In the absence of a second patient.

What to do? If Bertrand were the only patient known at this moment, it could take decades before another surfaced. Sequencing was becoming more common, but to find another patient, not only would sequencing exomes or genomes have to become common enough that another patient in another part of the world would get the test; not only would that team also have to know to prioritize the *NGLY1* gene out of all the possible altered genes in that patient's genome; but also, somehow there would have to be a way to connect that patient to Bertrand.

This was a huge problem. It really could take decades—and a whole lot of luck.

Matt wasn't prepared to wait for luck. Perhaps technology could provide an answer? He had a computer science blog. What if he wrote a blog post about Bertrand and mentioned his symptoms and signs? That would be a good start, but who would read it? He needed somehow to attain a broader reach. It occurred to him that if he could make one of his blog posts go viral, he could reach a lot of people. The post also had to rank in Google search results for all the right terms so that if another family were out there googling "lack of tears" or "dry eyes," they would land on his page. He thought for a while, then decided. For a job like that, you need Liam Neeson.

*"I found my son's killer.*
*It took three years.*
*But we did it."*

Matt began a blog post he titled "Hunting Down My Son's Killer." Below, Liam Neeson's piercing blue eyes stare out of a still-frame from the 2008 movie *Taken,* in which a former CIA agent fights to save his daughter from kidnappers. His gun is pointed directly at the viewer. His intensity calls to mind an iconic line from that movie, delivered in dulcet Irish tones studded with a razor-sharp edge: "I don't have money, but what I do have are a very particular set of skills."

Matt's story continues, beneath the Irishman's glare.

*I should clarify one point.*
*My son is very much alive.*
*Yet my wife Cristina and I have been found responsible for his death.*

Matt's blog went live on May 29, 2012. Then they waited.

. . .

Thousands of miles away and a few years earlier, Matt and Kristen Wilsey were preparing to move back to the West Coast from New York. Matt, originally from California, and Kristen, from Oklahoma, look like a Hollywood couple. Matt is tall, broad shouldered, with dark hair and a big smile while Kristen, blond, relaxed, and with an easy laugh, could pass for a Californian herself. They met at Stanford and got married at Stanford's Memorial Church less than a mile from the Stanford hospital where Matt was born. After graduating from Stanford Graduate School of Business, the Wilseys moved to New York for Matt's job. But their time on the East Coast wasn't to last. Kristen became pregnant, and as the pregnancy progressed, it became clear something wasn't quite right. Kristen had a prenatal screening that suggested a possible health problem with their baby, and although an amniocentesis (where a small sample of the fluid around the baby is analyzed) didn't reveal any abnormalities, they suspected something was wrong. They moved back to California to live near Matt's parents. Like most pregnant moms-to-be, Kristen was very sensitive to the baby's movements. On one particular day, at thirty-nine weeks and five days, she detected no movement at all, so she and Matt drove immediately to the hospital, where the baby's movements and heart rate and Kristen's uterine contractions were measured. A few minutes after the test began, chaos erupted. "They were literally ripping my clothes off," Kristen explained. "I was crashed, intubated, and taken for emergency cesarean section." Matt sat anxiously outside the operating room door in hospital scrubs.

Kristen woke up a few hours later to find out that she had given birth to a baby girl. She and Matt had been discussing names and had different first choices even the day before, but there was no hesitation when Matt was asked by the nurses for the name of their heroic little baby. He immediately offered up Kristen's preferred name: Grace.

At the beginning, Grace was lethargic and a little jaundiced (she had yellow skin, suggesting poor liver function), but neither of those things was too unusual for a newborn. What was less common was that a blood marker of infection was elevated. Yet over several weeks in the intensive care unit, no infection could be found. Eventually, she was allowed to leave the hospital.

Back home, life was not easy. Grace was not putting on weight the way most babies do. She was irritable, difficult to calm, and had trouble feeding. She had also started to twitch a little and turn her head sideways in an unusual fashion. Her liver function tests, still a little abnormal when she left the hospital, now yielded numbers more than ten times the normal range. Something was very wrong. Grace's muscle control was poor, and it became clear this was most likely a genetic disease.

Matt and Kristen attended the genetics clinic at Stanford, where they met Greg Enns, the director of biochemical genetics, and genetic counselor Julia Platt, someone predestined for genetic counseling by her childhood fascination with philosophy and ethics. She and Greg saw many patients with mystery diseases affecting the body's chemical reactions. "The stakes were so high," she told me, describing those days with Greg before she joined us in the Center for Inherited Cardiovascular Disease. "It really was a pursuit of passion." That pursuit drove the team to work sixty to eighty hours a week, toiling in the windowless basement of the hospital, compiling huge spreadsheets of data—every possible clue, every hypothesis, every test result enlisted to try to solve these mysteries. Julia recalled farmworkers from the California Central Valley who would take five-hour bus journeys to the hospital to present their ailing kids in clean, pressed clothes, at the clinic. "And you know, just feeling like you want to do well for all those families and being aware that many of them had no resources. It caused us to push ourselves very, very hard."

Matt and Kristen had ample financial resources and high-powered connections—and yet, in other ways, they were really no different from those farmworkers. Genetic disease knows no boundaries of culture or status. Everyone with an ill child navigates the same emotional journey.

Greg and Julia started working through the possibilities for Grace, carrying out a battery of gene tests. The approach was to methodically work down a list of possible diagnoses. At the time, most gene tests consisted of Sanger sequencing just a small number of genes, a pursuit frequently hampered by the insurance company's unwillingness to pay. Weeks of waiting followed each new test, and in Grace's case, those waiting periods were not rewarded with answers.

Matt had heard about genome sequencing. He knew it was still in the research domain. But if it was now possible to sequence a whole genome in one shot, why sequence only one gene at a time? Matt's background as a tech entrepreneur had taught him to move fast and gather all the data you can—even if you don't know what to do with it when you first get it. Matt connected with Baylor College of Medicine geneticist Huda Zoghbi, world renowned for discovering the cause of several devastating childhood diseases, and Richard Gibbs, an Australian geneticist, founding director of the Human Genome Sequencing Center at Baylor and leader of the team who sequenced Jim Watson's genome. Matt asked Huda and Richard if they would consider sequencing his family's genomes to look for an underlying cause of Grace's condition.

During this time, Matt's dad happened to be talking to his longtime friend, investor and former chairman of the board of trustees of Stanford, John Freidenrich. Hearing that the team at Baylor was about to embark on sequencing the Wilseys, John offered to connect Matt and Kristen to Mike Snyder and Atul Butte at Stanford. Atul was part of our team that attacked Steve Quake's genome in chapter 1, while Mike Snyder was Stanford's chair of genetics whom we met in chapter 5. Mike was at the time sequencing selected patients with undiagnosed disease, and he agreed to sequence Matt, Kristen, and Grace to try to find an answer. In another nod to his experience in business and technology, Matt elected not to tell the Baylor or Stanford teams that the other team was also working on the case: he wanted to see if they came up with the same answer independently.

The strategy appeared to work out well. After some time, both teams independently came up with the same top candidate: a gene called *SUPV3L1*. This convergence seemed exciting, but there was not universal agreement within the teams that this was the answer. As always, the challenge was not too few possible genes or variants; it was too many. One of Mike Snyder's postdocs, Mike Smith, had advocated for a different gene, but at the time, little was known about that gene and what was known couldn't be connected with Grace's symptoms, so it was pushed further down the list. Matt and Kristen joined Mike Snyder's lab meetings to discuss their daughter's sequencing results. They spent hours whiteboarding, going through genes that stood out in their analysis, piling up and interrogating the evidence for and against each one based on their intimate knowledge of Grace herself.

During this time, Matt and Kristen continued to fly around the country, taking Grace to as many experts as they possibly could. They were close friends with Brad Margus, a successful Florida businessman, whose sons both developed a devastating undiagnosed neurological disease. Brad took his sons to doctor after doctor, each stumped, until one day they met a doctor who just happened to have seen the exact condition before, immediately making the diagnosis of ataxia-telangiectasia. This was what drove Matt and Kristen to continue seeking out opinions, despite the cost and their fatigue—the possibility that out there was a doctor who had seen something like Grace's condition before. They were looking for that person to unravel the mystery.

In the Baylor team at the time was a Canadian postdoctoral scholar, Matthew Bainbridge. Richard Gibbs had walked past his desk one day with Matt Wilsey and suggested he take a look at Grace's genome. Bainbridge had written software to help prioritize gene variants to help solve undiagnosed disease. Straightaway, he fired up his software and set it to work.

The top gene, the one that both teams had initially been convinced was the culprit, was dismissed early on by laboratory experiments from Brett Graham at Baylor and Vamsi Mootha at Harvard. Bainbridge and the Wilseys decided on a new course. Instead of filtering the genes tightly to the specifics of Grace's medical presentation, they looked more broadly at all the genes disrupted in her genome. When Bainbridge examined that list, another gene target popped up, one that, although he didn't know it, had been on Stanford's original list. Bainbridge became increasingly convinced this could be the disease-causing gene. So he did the thing that every genomics professional does when they need to learn more about a new gene: he pulled up an internet browser and googled it. A moment later, he was surprised to find himself staring directly into the rather menacing blue eyes of Liam Neeson.

Matt Might's blog featuring Liam Neeson had gone viral. Within weeks, it was seen by millions of people. As well as the compelling hook of the Irish assassin, Matt made sure that the post contained keywords corresponding to the key elements of Bertrand's medical condition. There, in February 2013, scientist Matthew Bainbridge read about the unusual head movements, seizures, and staring. The similarities between Grace and Bertrand were startling (though there were differences too). Bainbridge immediately emailed Kristen Wilsey with several questions, including: Can Grace produce tears? One of the highly unusual aspects of Bertrand Might's presentation was that while he often cried out, he didn't make tears. This sign, alacrima, is incredibly rare.

Kristen received the email and replied the same day, answering Bainbridge's questions. "Yes, Grace can produce tears, though it is not often." The Wilseys had highlighted the lack of tear production to a handful of clinicians during the diagnostic journey but it didn't lead anywhere. Kristen added "I have only seen [tears] a handful of times in her three years."

When Bainbridge received the email, as he later told CNN, "it was one of those moments as a scientist where everything falls into place. You know, the one where you would run down the street yelling 'Eureka!' if that was something people actually said." Trembling a little bit, he forced himself not to tell anyone else at Baylor for twenty-four hours, waiting to see if he could poke any holes in his new hypothesis. Then he made a PowerPoint presentation and went to see who else he could convince.

One person he didn't have to convince was Mike Smith from Mike Snyder's lab. Smith had been arguing for *NGLY1* from the beginning. What was different now, however, was that everybody's original favored gene, *SUPV3L1,* had been discarded along with several other candidates. Also, now there were multiple patients with similar symptoms coming forward, including a family from Turkey whose physician saw the blog. All appeared to have mutations in *NGLY1.*

An enormous team of genome scientists along with a group of families working together, for years, across countries and continents, had defined a new condition and solved the case. The answer was *NGLY1.* The question was: What now?

# 10

# Next-Day Delivery

*There is no friendship, no love,*
*like that of the parent for the child.*

—HENRY WARD BEECHER

*One of the only ways*
*to get out of a tight box*
*is to invent your way out.*

—JEFF BEZOS, FOUNDER AND CEO, AMAZON.COM, INC.

The journey from diagnosing a disease to treating a disease can take many forms. The road is long and winding, but it begins with understanding clearly—at a molecular level—what biological process has gone wrong.

Proteins do the work of cells, and when they become old or damaged, those proteins need to be recycled. The product of the *NGLY1* gene is N-glycanase 1, a protein with an increasingly recognized diversity of roles inside the cell, including detaching sugars (glycans) from proteins so those proteins can be recycled. If this process is impaired, the consequences are serious. Proteins that should have been degraded build up. You have likely heard of some of the diseases caused by similar malfunctions in protein recycling, such as Parkinson's disease and cystic fibrosis. Add to that, now, NGLY1 disease.

NGLY1 removes sugars from proteins, which begs the question: Why would sugars be attached there to begin with? Well, it turns out it is useful for cells to be able to modify the function of proteins even after they have been made. This process is called *post-translational modification* (because it happens after the protein has been "translated" from the RNA). These modifications come in many flavors and involve extra chemical groups getting stuck on or taken off the proteins to regulate their function. In fact, proteins in the cell are constantly having these sugar groups and other chemical groups added or removed to allow them to respond to changes in the cellular environment. With sugars, this process is known as *glycosylation,* and it turns out the removal of these sugar groups (deglycosylation) depends critically on NGLY1.

To get a handle on such complex biology, Matt Might and Matt Wilsey undertook self-directed crash courses in biochemistry. As part of his course, Matt Might, Bertrand's father, went to see Lynne Wolfe, a specialist nurse at the Undiagnosed Diseases Program of the NIH and a colleague of Bill Gahl. Lynne took Matt painstakingly through all the biochemical pathways that involve NGLY1. He learned that to remove sugars, NGLY1 specifically breaks a bond between the protein and something called N-acetylglucosamine (fortunately shortened to GlcNAc, pronounced "glick-nack"). It appeared that a low level of GlcNAc in the cell was one consequence of Bertrand's and Grace's lack of NGLY1. A thought crept into his consciousness and took up residence. It gnawed away inside him. If GlcNAc was so important to deglycosylation, and it was low in patients with NGLY1 disease, perhaps you could treat the disease by replacing it somehow? But how could you possibly do that? He pulled up the Google search bar, hoping to read some scientific papers to get a clue as to how to raise the GlcNAc level, and was astounded to find . . . that he could buy it in tablet form on Amazon. Literally, he was staring at a bottle of it on his computer screen, right there, right then (some people take it as a supplement to help with joint pains). He moved the mouse, clicked Buy Now, and a few days later, through his front door, in a brown Amazon box, came a potential treatment for his son's devastating genetic disease.

But then Matt had second thoughts. This seemed too easy. So little was known about the role of NGLY1 and GlcNAc. Was it possible that this sugar supplement could harm Bertrand? Unlike for, say, an FDA-approved cholesterol drug, there was no clinical trial of ten thousand patients to look to demonstrating safety and efficacy. There was just the fact that malfunctioning NGLY1 caused Bertrand's disease and the likelihood that this contributed to a deficiency of the compound sitting in that cardboard box just delivered through his door.

Life intervened. Shortly after the box arrived, Bertrand was admitted to the hospital with pneumonia. This happened several times a year, and each time, Matt and Cristina worried that this would be the last time. So they decided: if he made it out alive this time, they were going to try the GlcNAc. "How do you forgive yourself as a parent if you had a potential treatment that might have helped your child and you never knew [if it worked]?" Matt said.

Now they had to work out a dosing regimen. The first thing Matt decided to do was give himself a huge dose. Large pharmaceutical companies, when testing a new drug for safety, give high doses to animals before they get anywhere near a patient. This was a supplement, a sugar, not a new drug but, still. Matt swallowed sixty grams in one day, about a whole bottle, and monitored himself for the next few days. Other than diarrhea, he noted no adverse effects. Using dosing guidelines from other similar supplements, he started six-year-old Bertrand on a pediatric dose of GlcNAc, and the waiting began.

Nothing much seemed to happen initially, good or bad. Matt wasn't sure how long any effect might take to develop, anyway. The disease had taken some time, during Bertrand's infancy, to accumulate its effect. Maybe the treatment might need months or years. Then on the third morning, Matt went into the kitchen to answer the call from Bertrand, who was crying out—hungry for breakfast. When he glanced at Bertrand, something looked different. Then Matt realized what it was. "I saw, for the first time in his life, tears streaming down his cheeks," he told me. As mentioned earlier, one of the defining features of NGLY1 disease is alacrima, a lack of tears. Yet here was Bertrand, crying his eyes out, three days after starting the GlcNAc. To make sure they weren't imagining it, Matt and Cristina took him back off the supplement. His tears dried up. They put him back on. They came back again. In fact, they have now repeated this multiple times with the same result. They even shipped his tears to an expert in glycosylation diseases, Hudson Freeze, director and professor of the Human Genetics Program at Sanford Burnham Prebys, to be analyzed for proteins. Later, Matt and Cristina also noticed that Bertrand's nighttime episodes (originally thought to be seizures) decreased when he was on the supplement and that he was sleeping more peacefully.

It appeared the risk was paying off, but Matt wanted to know more. To investigate the role of GlcNAc in NGLY1 disease further, he teamed up with Clement Chow of the University of Utah, an expert in fly genetics. Chow made a fruit fly model with the fly *NGLY1* gene disrupted. As expected, there was a decrease in GlcNAc in these flies. Further, only 18 percent of them survived to adulthood. When they supplemented the flies' diets with GlcNAc, starting at birth, 70 percent survived.

There were other tangible benefits to the new diagnosis, as well. One of these, for the Mights and Wilseys, was family planning. For many patients and families, this is a major benefit of understanding the genetic underpinnings of their disease. Some families find out, for example, that the genetic variant that caused their child's illness is not present in either parent—it arose for the first time in that child. This means that there is no increased chance of the disease recurring in a second child, something that can come as an incredible relief. For other families, understanding how the disease is passed from parent to child allows for intervention. The Mights went on to have two more children, and

Matt and Cristina wrote passionately about how knowing the cause of Bertrand's condition altered their experience. "The decision to conceive Victoria in the absence of a diagnosis for Bertrand was emotionally charged and difficult . . . As parents to one severely disabled child, we struggled with depression . . . At the depth of that depression, the thought of having another child came to represent a kind of hope. Yet, we knew that without an answer, there was a serious risk in conception." Kristen Wilsey also testifies to the power that a diagnosis can bring: "We felt incredibly relieved, hopeful. It meant that we could then have more children and screen for that."

Knowledge of the specific genetic variants that caused NGLY1 disease in Bertrand and Grace could be used to increase the chances of conceiving a healthy future sibling, thanks to a technique that has been around now for more than twenty years: in vitro fertilization.

The first successful pregnancy as a result of in vitro fertilization (IVF) led to the birth of infant Louise Brown on July 25, 1978, at Oldham General Hospital in Manchester, UK. The scientific breakthrough was reported by Patrick Steptoe and Robert Edwards in the journal *Nature,* and Edwards was later awarded the Nobel Prize in Physiology or Medicine for this work (since Steptoe had died, he was not eligible). At the time, Louise's conception and birth was front-page news. In vitro (literally, "in glass") fertilization involves extracting intact eggs from the ovary of the mother-to-be and combining them with sperm in a lab dish. Once an egg is fertilized, the embryo begins to divide, first into two cells, then four, then eight, then sixteen. At this stage, or just a little later, the embryo is a ball of cells, and one or more of the cells can be safely removed for genetic testing. Testing multiple embryos tilts the biological odds away from passing on the disease: only embryos found to lack the damaging genetic variants are implanted back in the mother, minimizing the chance the new baby will be affected.

This overall process, known as preimplantation genetic diagnosis (PGD), was first performed in 1989 for a baby at risk for cystic fibrosis.

It doesn't work for everyone, since there are many points at which the process can fail. In fact, PGD wasn't possible for Matt and Cristina, because the embryos lacking *NGLY1* variants were not healthy enough to be implanted. They did, however, go on to conceive a third child naturally and were able to confirm during the pregnancy that their new baby did not even carry a single faulty *NGLY1* variant. (This was done early in pregnancy via a technique called *chorionic villus sampling,* where a small sample of the fetal part of the placenta is taken for genetic testing.)

Meanwhile in California, Kristen and Matt's friends were asking how they could help Grace, so the couple set up a foundation to coordinate the attack on NGLY1 disease. The result, Grace Science Foundation, is the major global coordinated effort bringing together families with NGLY1 disease and researchers working on NGLY1 biology. The foundation supports 150 scientists in half a dozen countries. There are now more than fifty patients from thirty-eight different families known around the world. Through the foundation, Matt and Kristen communicate with NGLY1 families almost every day. They are cautious when they talk to newly diagnosed families. "Every new patient that we talk to, we try and set realistic expectations. It's unlikely their kids are ever going to be typical. But can they achieve some independence? Can they go from zero words to ten words? Can they go from using a G tube or being fed, to feeding themselves, and enjoy eating, which is such an important part of enjoying life?" Matt said. They certainly hope so.

The science continues. Grace Science Foundation has funded researcher Carolyn Bertozzi, a Stanford professor to study NRF1, a transcription factor whose job it is to control the expression of multiple genes. NRF1 controls another protein-recycling program in the cell called the *proteasome,* but its activity critically depends on NGLY1 removing a sugar group from it. Since NRF1 has such a profound effect in the cell, rescuing its function through some other means would appear to be a promising approach for improving the downstream consequences of loss of NGLY1.

Another pathway that appears to have potential for treating NGLY1 disease was discovered by Tadashi Suzuki, the scientist who originally discovered the *NGLY1* gene and who built the earliest experimental models. He found that the effects of NGLY1 deficiency could be rescued by inhibiting or genetically interfering with the expression of ENGase, another enzyme whose job it is to remove sugars from proteins. Pondering whether any already approved drugs had anti-ENGase activity, a team at the University of Utah built a 3D computer model of the enzyme, then scanned a database of drug compounds with known structures to see if any might be predicted to bind to ENGase. This is kind of like simulating a lock in a computer program, then having the program test the structure of thousands of keys in the lock, instead of having to actually make each key and try it. Their model found thirteen drugs that appeared to bind, and so the team ordered each one and tested each in a real test tube for its ability to inhibit ENGase. One class of drugs in particular was effective: over-the-counter antacid drugs called *proton pump inhibitors.* Two in particular, rabeprazole (AcipHex) and lansoprazole (Prevacid), appeared to bind particularly well.

Bertrand has now been on GlcNAc for some time and lansoprazole for two years. Recently, he spelled out his name on an Eyegaze computer (the mouse is controlled by the movement of the eyes). "I just freaked out," Matt said. "I was like, 'Oh my God, what is happening?'" When Matt would ask yes-no questions, Bertrand would now nod or shake his head.

"Oh my God," he said. "Bertrand can communicate."

And what about those tears? Recent work from Hudson Freeze may have finally provided the answer. Postdoctoral scholar Mitali Tambe found that whereas normal cells act like a sponge when placed in a water solution, filling up until they burst, when cells from children with an *NGLY1* mutation were placed in the water, nothing happened. It turned out the reason was that NGLY1 promotes the expression of water channels on the cell surface called *aquaporins.* Aquaporins are familiar to biologists, because they regulate the water content of many fluids in the body: sweat, the fluid that bathes the brain and spinal cord, saliva, and tears. Without NGLY1, aquaporin water channels are not expressed in the tear glands, and tear production is halted. Mystery solved.

Matt and Kristen have now run six annual NGLY1 meetings, bringing together NGLY1 families and NGLY1 scientists. To see the photos on their website is to reaffirm your faith in the power of collective energy and shared responsibility. It is one of the most striking demonstrations of the power that genome diagnostics offers. Sometimes people ask: Is knowing enough? They wonder: If there isn't a treatment, then what is the value of giving a name to a still-intractable disease? But the patients and families never ask this. "Not knowing is way worse than [just] having a disease," Kristen Wilsey told me. "You don't know what your prognosis is . . . You're just, like, stumbling along." Without the diagnosis, these families would not have come together, there would be no foundation, informed family planning would not be possible, and hundreds of scientists would not now be focused, together, on a cure.

# 11

# Hoofbeats in Central Park

*The things that make me different are the things that make me.*

—PIGLET, *Winnie-the-Pooh*

A. A. MILNE

*I am rare, and there is value in all rarity; therefore, I am valuable.*

—AUGUSTINE MANDINO

The Undiagnosed Diseases Network opened its doors on September 16, 2015, and over the first twenty months of operation, 1,519 patients were referred, and 601 patients were accepted for evaluation. In late 2018, we published a paper in *The New England Journal of Medicine* on our findings in which we described the complete evaluation of 382 patients. We were able to find a diagnosis in 35 percent of them, defining, in the process, 31 new syndromes and bringing that part of those patients' medical odysseys to a close. Beyond diagnosis, in four out of five cases where we found an answer, we were able to change something important about that patient's care: providing a specific therapy (like a drug or supplement) targeted to their condition, or reducing unnecessary diagnostic testing now that the diagnosis was clear. In many cases, we could also provide disease-specific genetic counseling to help with family planning. In this chapter, I'll tell you a few stories from some of our patients.

. . .

Take, for example, the case of seven-year-old Carson and five-year-old Chase. The first thing you're likely to notice about the brothers is their beautiful blond hair. Get a bit closer and you'll catch their sparkling blue eyes, and from that range, you can't miss their big smiles. These draw you in and hold your gaze so completely that you might not notice their wheelchairs.

We first met Danny and Nikki Miller and their disarming sons in 2017. At that point, neither child could speak, lift a cup to their mouths to drink, sit up unassisted, or walk. Their symptoms had developed gradually. Although Carson's motor milestones were a bit delayed during infancy, he did take a few steps at twelve months of age. This, however, was about as far as his motor skills developed. If anything, he seemed to go a little backward in his development over the next year. His little brother, Chase, was born around that time, and the joy of his birth was tinged with worry. Something really wasn't quite right with his big brother's development. Would this also be true for Chase?

Danny and Nikki had taken Carson to behavioral specialists, then to developmental specialists, and then to one neurologist after another, but no one had any answers. They were eventually referred to their regional specialist center to see experts in rare neurological disease. There, with Carson about sixteen months old, they were given a diagnosis of cerebral palsy. This broad label (literally meaning "brain paralysis") didn't reveal anything, however, about what had actually caused Carson's problem or how best to treat it. Nor could it help them determine if any future children of theirs might be at risk. They needed more information.

Their worries did not diminish a year later when Chase, too, began to miss developmental milestones. It was clear that both boys were developing normally from an intellectual standpoint but that both were suffering from a severe movement disorder. Both boys' MRIs, in fact, showed lesions in the brain's basal ganglia, which are responsible for coordinating smooth, intentional movements. With both boys affected, there was now little doubt that this was a genetic disease. Genetic testing began through their local doctors. Initially, small numbers of genes were sequenced, and eventually, Carson and Chase underwent full exome sequencing of all twenty thousand of their genes, not just once but twice. But still, answers were elusive. For three more years, Danny and Nikki were consumed with caring for the boys while trying to figure this out.

From late night until the early hours, after the boys were asleep, Danny and Nikki pored over the accumulating volumes of medical notes and searched every corner of the internet, trying to understand the possible diagnoses and what each one in turn might mean for their family. As Danny put it, "There was a lot of uncertainty, a lot of angst, a lot of sleepless nights. I mean, I would be up at two in the morning, researching for several hours at a time, because I couldn't sleep." That was when, in 2016, they first heard about the Undiagnosed Diseases Network (UDN). With help from their fifth neurologist, they submitted an application.

When we read about the severe, early-onset neurological symptoms in two young brothers born to healthy parents, it was an easy decision to accept the boys into the network. We then got down to the work of trying to solve their case.

As we do in almost every case in the network, before we even met the family, we sequenced their genomes. True, Carson's and Chase's prior exome sequencing (which looked at all twenty thousand of their protein-coding genes) had not found any answers. But whole-genome sequencing has some distinct advantages: there is more accurate representation of the entirety of each gene, whereas exome sequencing can be a bit patchy especially at the beginning and end of the genes where the regulatory elements live (the all-important switches that turn genes on and off). There is also better detection of large disruptions in the genome, such as large chunks of DNA that have been deleted or inserted where they shouldn't be.

In this case, we elected to sequence the family "quartet" in order to compare the boys with their parents. While that was being done,

Jonathan Bernstein, Stanford's chief of medical genetics, examined the boys in clinic to better understand the nature of their motor delay. Jennefer Kohler, one of our genetic counselors, joined him to review with Danny and Nikki where things were with the genetic evaluation.

When the results came back in March 2018, the power of having full genome sequencing on all four family members became apparent. Because Danny and Nikki did not suffer from the condition, our search focused only on genes that we knew could cause a motor delay, where each boy had received one disruptive gene variant from each parent. Since Danny and Nikki developed normally, it was likely that a combination of two variants was what caused the disease. It appeared Carson and Chase had both received two disruptive variants in the gene *MECR*. The exome sequencing they previously had undergone didn't detect one of the variants, since it was in a regulatory region not covered by the exome.

Jennefer put together the evidence for and against this gene, as well as the other top candidates, in a slideshow. Then our whole team sat around a conference room table and interrogated the evidence. First, we focused on the gene itself. What did we know about it? As it turns out, the answer was "Not much." Where did the protein coded for by this gene live? That one seemed clear from prior research—it lived in the mitochondria, the energy-producing powerhouses of the cell. What did the protein do there? It seemed to be part of an energy pathway. What other proteins did it interact with? We just didn't know.

After looking for evidence that the gene might be capable of causing the boys' disease, we then looked at the specific variants themselves. How disruptive were they to the gene? In this case, one was very disruptive, because it changed the way the gene message was processed while the other was less clearly disruptive, since it seemed to affect only the regulatory region. Did our computer models "guess" that these variants would be damaging or harmful? Most models predicted one or both variants would significantly disrupt the protein's function.

Jennefer had dived headlong into this evidence. One key paper described patients with variants in the *MECR* gene and had been published in the *American Journal of Human Genetics* in 2016. In it, seven patients from five families were described. This looked like a promising lead, but Jennefer had been here many times before. Often, childhood diseases cause similar brain and nerve problems, and especially given one variant in *MECR* was in a regulatory region, it was hard to be sure this was a match. But as she began to read the paper, she noticed something unusual: all the children in the paper appeared to have speech and motor delay but preserved cognition (intellect). She remembered seeing that phrase somewhere before. She pulled up the clinical notes on Carson and Chase from Stanford geneticist and child neurologist Maura Ruzhnikov. And there it was, that exact phrase: "speech and motor delay with preserved cognition." It was starting to look very promising indeed. She read further and found that some of the patients described in this paper also shared brain scan features with Carson and Chase, changes in the basal ganglia, the part of the brain responsible for movement. It seemed very likely that if both of the variants we identified could be shown to be damaging, then this had to be the answer. For one variant, it was obvious it was damaging from the genetic change itself. But for the variant in the regulatory region, only an experiment could confirm that it disrupted the protein—and for that, we needed cells with that specific variant in them. Our laboratory team cultured skin cells taken from the boys and then sequenced all the RNA messages looking to see if the *MECR* message was disrupted. Shortly thereafter, the jury returned a verdict: both variants looked damaging, and Carson and Chase became the eighth and ninth patients in the world to be diagnosed with MEPAN syndrome (a lot easier to say than *mitochondrial enoyl CoA reductase protein-associated neurodegeneration*).

The rest of the team got to work on using this information to find a way to treat the boys. The protein made by the *MECR* gene is part of a complicated energy-producing pathway inside the mitochondria.

Bear with me here for a moment, and I'll explain how it works. Imagine a toy factory making dolls. Each worker in line is delivered the doll in some state of assembly. They add their parts to it, then push it on down the line to the next worker. This is how important energy molecules in cells are made. Imagine now that one of the workers is off sick (this is the disrupted gene). Let's say it is the worker who takes the body and adds the arms and legs. Several things happen as a consequence of that worker being unavailable. First, the next worker down the line, who is supposed to add clothes and shoes, receives only the body and can't do their job. This leads to a buildup of limbless dolls that can't progress down the assembly line. There is also a buildup of unattached arms and legs. Finally, production of the finished dolls comes to a complete standstill. All these factory line issues can separately and together contribute to disease. The best answer, clearly, would be to call in an experienced, qualified replacement worker. Then the whole pipeline would start up again. This would be the equivalent of putting a normal gene back into exactly the right cells at exactly the right time. This is the holy grail of gene therapy, but it remains, unfortunately, very difficult to achieve (see chapter 19). So what if that's not possible? What if, instead, you could bypass the problem by delivering some extra dolls, complete with arms and legs, ready to be clothed before rolling on down the factory line to the boxing department?

In the case of MEPAN syndrome, those extra dolls would be a molecule called *alpha-lipoic acid.* Adding that would bypass the boy's genetic problem by supplying something their faulty *MECR* gene could not produce. And as luck would have it, you can buy alpha-lipoic acid on Amazon for $16.95 a bottle (some people use it as a supplement to help with nerve problems from diabetes).

Carson and Chase started taking alpha-lipoic acid as well as medium-chain triglyceride (MCT) oil to help replace the missing parts on the factory line. Amazingly, after a few months, Danny and Nikki started to see a difference. Prior to the diagnosis, both boys had been losing ground from a motor skills perspective. In his second year of life, Carson had lost the ability to sit independently and to take a few steps. Chase had been crawling early in life, but since the time he was about a year old, he was gradually losing that ability. Since they've been taking the supplements, their decline stabilized, and their neurologist has actually seen improvement. "That's a huge thing for us," Danny added. In the spring of 2019, after the paper describing their boys' diagnosis was published, Danny and his family were kind enough to share their story. They appeared on *CBS This Morning* and on NPR's *Morning Edition* to give a picture of their family life and what the Undiagnosed Diseases Network meant to them. Danny told me, "The UDN is an amazing resource for undiagnosed families with nowhere else to go for answers. There's no other program like it—it's an opportunity to see some of the best researchers and clinicians in the world, and it's available to anyone, regardless of their means. That's pretty unique in today's health care landscape. They provide hope for rare disease families that sometimes have none."

Having a diagnosis allowed Danny to start to connect with other families suffering from MEPAN syndrome, as well as scientists around the world working on this metabolic pathway. He also started to raise money to push the research further. In early 2019, he founded the MEPAN Foundation. One of the central goals of the foundation is to better connect families with MEPAN syndrome. For Carson and Chase, we were fortunate to find a scientific paper and connect the dots of similar movement disorders and brain scan findings in other families to help make the diagnosis. But to rely on the standard publication process is to hope that some geneticist somewhere happened upon enough patients with a new disease that she decided to write them up, that the lengthy editorial process led to the paper actually being published, and that the paper actually was spotted by other doctors with similar patients across the world. It is simply not a good enough solution for connecting patients, families, and investigators when people's lives are at stake. Fortunately, we now have better approaches.

One of these is called Matchmaker Exchange, a forum led by Heidi Rehm, a professor at Harvard and the Broad Institute. The exchange provides a common language and central gathering place for information about rare disease cases, allowing secure sharing of information (such as patient symptoms, signs, blood test results, and genome sequencing results) that can be critical to making diagnoses. Now we regularly receive messages through this forum and others, asking if we have ever seen a patient with a certain set of symptoms or a gene thought to cause a set of symptoms. Eventually, this will happen in an automated manner and large computers will crunch though large amounts of data every day and every night, connecting the dots.

Several key features of the UDN allowed us to find a diagnosis for Carson and Chase and to solve cases like theirs, where others had failed. One is that because of NIH support, we are not limited to using tests covered by a patient's insurance (as is the case in normal medical practice). Techniques still in the realm of research, like genome sequencing or even RNA sequencing, could therefore routinely be used. What's more, we were always able to sequence additional family members beyond the patient referred—harnessing the tremendous power of comparing affected and unaffected family members to narrow our "suspect" gene list significantly (as we first did in the West family in chapter 5).

As discussed in chapter 7, we are occasionally asked whether all this extra testing won't just bankrupt the health care system. In truth, genomic medicine actually saves money—which is worth emphasizing as we try to help more families. Patients with undiagnosed disease are among the largest contributors to health care expenditure. Whereas genome sequencing currently costs under $1000, it costs $10,000 to spend one day in an intensive care unit—which is where undiagnosed patients all too frequently end up. When we tallied up the medical costs for a group of UDN patients who received a diagnosis, the average cost of care before coming to the UDN was over $300,000. By contrast, the average cost of their UDN evaluation was under $20,000. The implication is clear: if we had been able to offer genome sequencing to some of these patients sooner, their health care costs may have been reduced by as much as 94 percent. And this is before you factor in the less measurable benefits, like sparing families emotional anguish. Keep in mind, too, that for many developmental illnesses, there is a critical window for treatment where a replacement therapy could be transformative for life. If a diagnosis is made early, treatment can be started early with potentially dramatic changes in lifetime burden of disease and need for rehabilitation.

Although 74 percent of our cases are solved by genomic sequencing, more than 10 percent of cases are solved by combing through the charts, looking carefully for clues, following leads, and bringing together experts. We think of this as the Sherlock Holmes approach.

In 2016, for example, when we first met Robert Leathers at Stanford's branch of the UDN, he was a seventy-year-old man, who, for some eight years, had been suffering recurrent drenching fevers, troublesome skin rashes, and severe muscle pains. The fevers had at first been attributed to a medication he was taking. However, when he stopped the medication, nothing changed. Another hypothesis was that he might have an unusual form of tuberculosis, so he had been treated with antibiotics for tuberculosis for many months, but there was still no change. Over the next few years, on the possibility that an overactive immune response was at fault, he also tried three medications designed to broadly suppress his immune system. There was minimal improvement. After five years of these symptoms, he sought a third opinion at the Mayo Clinic in Rochester, Minnesota, where they carried out CT scans, ultrasounds, MRI scans, multiple blood tests, and genetic tests for rare fever syndromes, such as the exotically named familial Mediterranean fever. All were uninformative. He all but gave up hope of ever finding an answer. Three years later, during a visit to the neurology department at Stanford for a fourth opinion, he was referred to our Center for Undiagnosed Diseases.

He sat before us now, desperate to know what was wrong with him. There weren't many clues in his case to guide us. He did have abnormally high levels of a protein derived from antibodies in his blood, suggesting that his immune system was activated, but beyond that, there was very little to go on. Internist Jason Hom and Matt Wheeler took the lead in gathering all his prior records, reviewing all his charts, and choosing specialists in blood, skin, and joint diseases to help us try to find an answer. They also asked our infectious diseases team to see him, because the most common cause of a recurrent fever and sweating is infection.

Our infectious diseases expert, Shanthi Kappagoda, was the first to put together the unusual combination of a raised, red, and itchy rash, the antibody protein in the blood, and long-term intermittent fevers. Shanthi knew that familial Mediterranean fever was a possibility (even with a negative gene test), but it is unusual to suffer that for the first time in your sixties. Muckle-Wells syndrome also causes fevers, but it is unusual to see it without deafness. The same goes for cryopyrin-associated periodic syndromes (CAPS), another set of rare diseases that present with joint pain, rashes, and fever. In the same way, several other possibilities were raised and then discarded.

One rare condition, however, seemed to be a better fit with the data: a syndrome first described in 1972 by the French dermatologist Liliane Schnitzler. Only about 160 cases of Schnitzler syndrome have ever been reported. The disorder is classified as *auto-inflammatory,* meaning that the body's natural response to injury—inflammation—which causes the swelling and redness after, say, a cut or scrape, is turned on when it shouldn't be.

How to test this hypothesis? Well, first, did we have evidence of inflammation? Yes. Our patient had a raised overall white count in the blood and a raised marker of inflammation called *C-reactive protein.* If indeed the culprit was Schnitzler syndrome, we knew, the skin should show invasion of a particular kind of white blood cell called a *neutrophil.* Did it? Indeed, those cells were found when his inflamed skin was examined under the microscope. One step closer. What other evidence could we look for to confirm we were on the right track? Another key feature of Schnitzler syndrome is the class of abnormal antibody proliferating in the blood; usually, it is one known as an *M antibody* (this is the largest antibody and the first one to appear in response to an infection). Yet it appeared the abnormal antibody protein in Robert's blood was mostly *G antibody* (the most common antibody, making up three-quarters of the antibodies in your body), which is actually *not* typical for Schnitzler syndrome. On deeper review of the medical literature, however, we learned that a tiny number of patients with Schnitzler syndrome had indeed presented with this G antibody instead of the M antibody. Inching closer still. Finally, what about pertinent negatives, any curious incidents of dogs that "didn't bark in the nighttime"? In this case, the absence of any major joint problems allowed us to rule out many other possible diagnoses that we would normally suspect when presented with a patient with rash and fevers. Schnitzler syndrome: the diagnosis was made (and not a genome in sight).

An important aspect of making a diagnosis, of course, is being able to target therapy. Enter our chief of rheumatology, Cornelia Weyand, who has dedicated her career to understanding how the immune system ages and how this leads to inflammatory disease. Her group defined the mechanisms by which inflammatory cells invade tissue and then manage to stay there, causing problems such as redness and swelling. She recognized these inflammatory features in Robert and started to form a therapeutic plan. For patients with overactive immune systems, we often use medications like steroids that suppress every aspect of immune system function (Robert had tried this before). As you might expect, however, using a sledgehammer to hit a small nail is effective in driving the nail but can cause significant collateral damage. These include side effects ranging from weight gain and bloating to skin changes, tendon rupture, and even depression and mood swings. The good news was that, having identified his condition as auto-inflammatory, we could focus his therapy more precisely on the root cause—inflammation—our understanding of which has advanced dramatically over the last few decades. One important regulator of inflammation is an immune system hormone called *interleukin-1.* It has been targeted by many drug companies over the last few decades as a master regulator of inflammation. In fact, when Robert was put on an interleukin-1 blocker, there was an improvement in his symptoms; however, by the time he was referred to the network, he was again having fevers despite this therapy.

In 2012, a report was published in *The Journal of Allergy and Clinical Immunology* of three patients with Schnitzler syndrome who had sustained success with inhibition of a closely related hormone, interleukin-6. We discussed this finding with Robert's doctors, and switched him to this blocker. I caught up with Robert on the phone late in 2019 to see how he was doing. The medications were working well. I asked if he felt, after eight years of toiling without an answer, the Undiagnosed Diseases Network had helped his case. "Absolutely," he said without hesitation.

Robert's case teaches us that sometimes the traditional tools of observation, discussion, and inference can take us all the way to an answer, without the genome. Indeed, especially if your problem appears to emanate from the immune system—a system that adapts throughout life according to your exposures, infections, and vaccinations, then we wouldn't expect that sequencing the genome you are born with will be very helpful in finding a solution. Yet immune cells manage to adapt to these exposures and attacks by rearranging certain areas of their own genomes. This allows them to respond to and shut down an almost infinite number of known and unknown attackers. If we could isolate those specific immune cells and then sequence the specific areas of their genomes that we know are rearranged to create this adaptive immunity, we could open a new door to the understanding of immunological disease. In fact, this is exactly what is happening. In some patients, we can now sequence the immune system to start to understand exactly how their immune systems have adapted. Soon, we will be able to leverage this information to target precise therapies at increasingly well-defined immunological problems.

In 2014, a four-year-old girl, Allyssa Lawson, was referred to our center. She had been born with unusual facial features and was suffering from joint problems. When our pediatric geneticist Jonathan Bernstein examined her, he noted Allyssa's loose joints, a curving of her spine, protrusion of her chest, and flat feet, all of which suggested a rare connective tissue disease called *Marfan syndrome*. However, confusingly, she also had features that were not part of Marfan syndrome: thin hair and eyebrows and a pear-shaped nose sometimes seen in an even rarer condition called trichorhinophalangeal syndrome (TRPS1). He was puzzled. This was an unusual combination of features. There was also the fact that Marfan syndrome is always associated with tall stature, while TRPS1 is always associated with short stature; Allyssa's height was normal. Was that a vote against each of those diagnoses? A thought occurred to him. Could the "tall" effect from Marfan syndrome be canceling out the "short" effect from TRPS1, making her height normal and confusing all the diagnosticians? Could she possibly have two diagnoses?

In medicine, we often implore our students to apply Occam's razor to their clinical reasoning: if two rare symptoms appear to be occurring in the same patient, it is much more likely they are connected to one cause than two. The philosophers state this more elegantly: all other things being equal, simpler explanations are more likely. And although, as humans, we are attuned to detect rare or unusual things, in fact common things are common. Or as one version of a phrase famously attributed to Theodore Woodward of the University of Maryland in the 1940s states, "If you hear hoofbeats in Central Park, think horses, not zebras."

Given that Marfan syndrome occurs in one in five thousand people and TRPS1 in one in a million, the chances of the same individual having both these conditions is somewhere around one in five billion. To put this in perspective, the odds of being struck by lightning in your lifetime are only one in three thousand, while the odds of winning the U.S. Powerball lottery are currently one in three hundred million.

However, in the field of rare disease medicine, everywhere we look, we see zebras. In Allyssa's genome, our genetic counselor Diane Zastrow found suspicious variants in genes for both Marfan syndrome *and* TRPS1. With a one in five billion chance of this happening and around seven billion people on the planet, as my colleague Matthew Wheeler put it, "She's basically unique in the world."

With these two new diagnoses came a clear plan of action. Allyssa is now followed by medical specialists for each condition and undergoes the recommended screening for her aorta, the large blood vessel that comes out of her heart, that can be enlarged in patients with Marfan syndrome.

Allyssa is now a bright and happy second grader. She loves horses and dragons and being a big sister to her little sister, Carlee. She is passionate about animals and recycling and has hopes of saving the planet one small change at a time. As her mom told me, "Allyssa understands her medical limitations but doesn't let them stop her from living her best life."

. . .

Sometimes, the diagnosis can be made by examining the patient, searching through records, hunting for clues, and thinking hard; sometimes, it can be made by sequencing the patient's genome and searching through the genomic clues. Sometimes, however, all of that is just not enough.

Imagine, for a moment, that feeling you get when you have worked out as hard as you possibly can: your muscles ache, you feel sick. Your body is overwhelmed. A lot of that is caused by a buildup of acid. Fortunately, it's temporary. As you cool down from your exertion, your body starts to recover and fix your acid level, and you soon start to feel better. Now, however, imagine that your body started producing excessive amounts of acid when you weren't exercising—perhaps in response to a stressful day or when you caught a minor cold. Imagine now that it doesn't ebb away as it does when you recover from exercise. In fact, imagine such a severe buildup of acid that it could tip you into a coma-like state requiring the intensive care unit. That is what Anahi, a little girl who we first met at age six, was living through.

Normally, these metabolic derangements are caused by variants in the genes that code for the energy-producing apparatus in our mitochondria, the powerhouses in every living cell that produce energy molecules called *ATP.* Inside each mitochondrion, there are five well-recognized engines working on this ATP production, called complexes I–V. Often complexes I–IV can be mutated in a way that causes disease (complex V was generally felt to be "too critical" for someone to survive with a mutation). Perhaps the problem lay with one of them? If these engines were misfiring, then the cell has to make energy another way, and the byproduct from that process is acid. Yet when we looked at the genes for complexes I–IV, as well as measured their actual energy output in Anahi's cells, nothing was abnormal. That was unexpected. How could we explain Anahi's condition if her mitochondria were working just fine?

Genome sequencing provided the first key insight, thanks to the fact that it sequences all of a person's genes and not just the well-known ones. In Anahi, Ph.D. curator Dianna Fisk and genetic counselor Megan Grove found a variant in the gene *ATP5F1D* for the critical fifth engine for ATP production inside mitochondria. In fact, both Anahi's copies were affected by the same variant. But despite its role in building the fifth engine, this specific gene had never been associated with disease before. So how do we prove that this is the answer? How do we move from candidate gene to new syndrome?

In the world of rare disease, getting the word out can make all the difference. In 2015, Liliana Fernandez, a coordinator and geneticist on our team originally from Colombia, and genetic counselor Jennefer Kohler met Anahi and her mom for the first time. A year later, they presented a poster on her case at the American Society of Human Genetics meeting in Vancouver. As it happened, a researcher from Newcastle in the UK, Robert Taylor, had been studying a local patient with a similar presentation and had also zeroed in on this gene. He was perusing the conference program when he was taken aback to see the name of the gene he had been working on in front of him on the page. He immediately contacted Anahi's doctor, who connected him with our team. Jennefer and Robert arranged to meet in front of her poster, where they discussed the remarkable similarity between the two cases, including the fact that both children had been born to healthy parents. In each patient, both copies of the gene were affected. Suddenly, each team appeared to have found their second case. But this alone was not enough proof. We also needed to demonstrate that the variants in this gene actually decreased energy production by this fifth engine, the so-called complex V.

Enter the laboratory component of the UDN, where we take our hypotheses to the lab and design experiments to try to solve the mystery. The goal here was to directly measure energy production from each of the five complexes in skin cells derived from our two similarly affected patients and to show that it was specifically the energy production from complex V that was decreased. This was done by blocking each of the complexes in turn in the mitochondria from the patients' cells and measuring the energy output. Indeed, it was dramatically decreased in both children. Furthermore, the mitochondria within each patient's skin cells looked misshapen under the microscope, but they were also sparse and difficult to study, because skin cells don't need too many mitochondria. Heart cells, however, are packed with mitochondria to enable them to beat. So to examine the sickly mitochondria in greater detail, we transformed some of those skin cells from the patients into stem cells (cells that can turn into any kind of cell in the body). Then using a specific "heart cell" recipe, we turned them into heart cells that began to beat spontaneously in the lab dish. We created, in effect, mini patient hearts. Since hearts beat constantly, they are very dependent on mitochondria to provide energy for muscle contraction—and the mitochondria in the heart cells from these patients looked particularly unusual and misshapen. Together with the fact their energy production was impaired, the story was beginning to come together.

These were cells in a dish, however. Would Anahi's altered complex V actually cause problems in a living organism? Again, we drew on the unique powers of the UDN to answer that question. Within the UDN is a specialized group at Baylor College of Medicine in Texas who help solve cases by modeling genetic diseases in fruit flies. Flies are useful models in genetics because they breed quickly and genetic modifications can be made easily. Of course, they are evolutionarily far from humans, but they are a step up from cells in a dish, and to study the basic function of very fundamental genes, they can be very helpful. To investigate our patients' variants, Hugo Bellen's group at Baylor started with a cool trick: they generated flies in which the fly version of the human gene could be turned on or off at will. After that, they set about adding copies of the human gene to the flies: either the normal human gene or the human gene mutated exactly as it was in our patients. When they turned the fly gene off completely, the fly larvae did not develop at all. So it was clear that this gene was critical for the living organism. Next, they turned the fly gene off only in the brain and neural tissue. Those flies survived a little longer, but they still died very young. Next, they turned off the fly gene and, at the same time, turned on a normal human gene. Those flies switched to using the human gene to produce energy and survived well. So now we knew quite a lot about this gene in general: the normal human gene could "rescue" the fly when its own gene was turned off. The question now was, could a mutated human gene (Anahi's version, or one from the child in the UK) perform the same rescue? If yes, that would suggest that the mutated human gene was actually functional, so perhaps not the answer to our patient's disease. Hugo's group turned off the fly gene and turned on one copy of each patient's mutated gene in different groups of flies. All the flies died, providing clear proof that the patients' mutations disrupted the gene's function in a living organism.

Matt Wheeler pulled all this work together and, with colleagues in the UK who had studied the other patient, reported the discovery of a new disease in the *American Journal of Human Genetics* in early 2018. The disease now has an official name: mitochondrial complex V (ATP synthase) deficiency, nuclear type 5—or MC5DN5 for short. As Anahi's mother told the *San Francisco Chronicle,* "At least the doctors know exactly what she has, and maybe they can treat her better now." In fact, although we do not yet have highly precise or effective treatments for mitochondrial disease, identifying the causative gene and mechanism meant a range of treatments in the form of supplements and dietary and lifestyle advice could be given to these two young patients, separated by an ocean. In the future, their coming together to define a new disease could help many more patients find answers and hope.

We have so many ways to reach a diagnosis, but despite our efforts, the majority of cases remain unsolved. For the 65 percent of patients who remain undiagnosed, we do not lose hope. Science continues to move forward. Every month, every year, new findings are reported. Sometimes a diagnosis comes from simply taking a fresh look at the already existing sequencing data in light of newly reported gene-disease associations.

We continue to play disease detectives, investigating new cases with a view to finding diagnoses and, someday, cures. In 2018, the NIH expanded the network to twelve clinical sites. There are now undiagnosed diseases programs in many other countries, and Bill Gahl chairs an annual meeting for the UDN International group. In 2019, Gina Kolata featured some of our patients in an article for *The New York Times.* It appeared in the Week in Good News section. Recently, I asked Bill whether he foresaw this kind of impact when he first accepted the offer to open an Undiagnosed Diseases Clinic at the NIH. He let out one of his characteristically hearty laughs. "I would never have expected it to go anywhere," he said. We can all be thankful that his persistence and the inventiveness and resilience of the teams and families around the world have led us to the brink of ending medical odysseys for almost half of our patients. There is more work to be done, but even Sherlock Holmes himself might have offered some grudging appreciation for that.

Part III

# AFFAIRS OF THE HEART

# 12

# Whisky à Go-Go

*Dancers don't need wings to fly.*

—RAVI NATHANI

*Hearts will never be practical*
*until they can be made unbreakable.*

—*The Wizard of Oz*

L. FRANK BAUM

We first got to know Jazlene before she was even born. At thirty-six weeks of gestation, still inside her mother, her obstetrician had noticed that her heart was beating around 70 beats per minute. This is a very reasonable heart rate for an adult sitting down reading a book, but not remotely close to the 140 beats per minute that would be normal for a six-pound baby swimming around inside her mother's womb.

Her low heart rate led Jazlene's doctors to diagnose some form of heart block, in which electrical signals were hitting an unexpected stop sign in their route through the heart. So she and her mother were transferred to Stanford's Lucile Packard Children's Hospital. There, it was decided to carry out an emergent cesarean delivery. Jazlene was mature enough to survive outside her mother but clearly needed immediate help from a cardiology specialist. The procedure went well, and as a new citizen of the outside world, she had the first of many medical tests, an electrocardiogram. That ECG was carried out by Anne

Dubin, a world-renowned electrician for kids' hearts. Anne is straight up, down to earth, high energy, and brilliant, with a knack for making complicated concepts sound like, well, child's play. In other words, exactly who you want if your newborn appears to be suffering from a mysterious electrical problem of the heart. Anne's ECG made it clear that Jazlene was not suffering from a block in the traditional sense. Rather, she was suffering from a particularly severe form of long QT syndrome, a rare disease in which the electrical resetting of the heart after each heartbeat is prolonged. The real issue was not with this prolonged resetting or even with the resulting low heart rate but rather that the electrical abnormality, the lengthened QT, could lead to potentially fatal heart rhythms. Such life-threatening rhythms are especially likely to be triggered in situations of stress. And let's face it, there aren't many situations more stressful than being born. In a cesarean delivery, you are pulled in a matter of seconds from that warm, watery cocoon that has enveloped you since you first became conscious and are exposed to a cold, unfamiliar outside world. Suddenly, you have to breathe, something you have never had to do before. The dull, muffled sounds that were so distant through the barrier of amniotic fluid are now loud to your sensitive ears. There are unfamiliar voices and lots of beeping, shrill alarms. The one voice you recognize above all others is the one you now don't hear. Then you are being handled, moved horizontally and vertically in jerky motions you are not accustomed to. You are being rubbed, blanketed, the sensations inconceivably foreign. In response to this stress, on her first day in the outside world, Jazlene's heart stopped beating multiple times. Anne's fingers were not only on her pulse that day but also on her thin chest, doing CPR, pushing blood around her tiny body, keeping her alive. Indeed, a rapidly growing team of experts worked round the clock to figure out what specifically was wrong—and more importantly, how to save her.

. . .

The story of Jazlene's condition begins in the 1850s at an institute for deaf children in Leipzig, Germany, when German obstetrician Friedrich Ludwig Meissner was called to investigate a startling and traumatic event. A young girl, both deaf and "mute," had stolen another child's belongings, and the head of the institution, a Mr. Reich, had summoned the girl in front of the other children. His idea was to give her the opportunity to confess. As Meissner put it, "Seeing the disapproval of the teacher whom she adored made her feel pain and regret." But the lesson did not go as planned. After being called out, without any punishment yet administered, she collapsed and died. Worse yet, when Dr. Meissner went to break the bad news to the girl's parents, he learned something that deepened the tragedy further. This was not the first time her parents had received such news: two of their other children had also died suddenly in similar circumstances, one after a fright and another during a fit of rage. Meissner wrote up the case in 1856, describing the coexistence of childhood deafness and a tendency to collapse suddenly and die. This was decades before the ECG was invented and more than a century before two Norwegian doctors first characterized the heart condition we now know as long QT syndrome.

Anton Jervell and Fred Lange-Nielsen were physician researchers at the University of Oslo, in the cultural capital city of Norway. In the 1950s, they had come across a family with that same curious combination of deafness and sudden death. When they studied the ECGs of the six surviving siblings, four of whom were deaf and had a history of fainting, they saw a prolonged QT interval in just those four. They also saw extra heartbeats and short runs of an uncoordinated-looking rhythm that seemed plausibly connected to the sudden collapse. Jervell came to dedicate many years of his life to this disease that he called *surdocardiac syndrome* (*surdo* is the Latin for "deaf") and what the rest of us now call Jervell Lange-Nielsen syndrome.

In each of these families, notably, the children's parents were unaffected, having normal heart rates and normal hearing. The Jervell

Lange-Nielsen syndrome was soon confirmed to be inherited in a manner known as *autosomal recessive,* meaning that both parents contribute a disease-causing genetic variant for the syndrome to manifest when those two variants meet in the child.

A few years later, in 1964, an Irish pediatrician and cardiologist by the name of Owen Conor Ward was referred a mysterious case of a six-year-old girl with repeated attacks of fainting but normal hearing. At sixteen months of age, she had collapsed while running. The attacks then became increasingly frequent until, at three years old, she was having them every few days. At times, they were associated with convulsions. Even hospital admissions had failed to find a cause, since no episode had occurred during any of those admissions. She was a true medical mystery. A psychological cause had not been ruled out. Hoping he could help, Dr. Ward admitted the young girl to his hospital, where he tried to re-create her symptoms by having her run around the hospital grounds wearing ECG electrodes. During one of her runs, she suddenly collapsed. The team rushed to perform an ECG, and instead of the regular series of P-QRS-T waves that characterize every heartbeat on a normal ECG, the tracing showed an ominous, wide, twisting rhythm. This rhythm appeared similar to ventricular fibrillation (that disorganized electrical rhythm where the heart vibrates instead of beating, pumping no blood) but was different in one important way: the irregular waves seemed to twist back and forth on the sheet of paper without a central axis to anchor them down. In 1966, the French cardiologist François Dessertenne labeled these irregular waves *torsades de pointes,* meaning, literally, "twisting of the points." Yet the romanticism of the French name belies the ugliness of the rhythm described and its frequent role as harbinger of death.

The girl recovered, but Dr. Ward was puzzled by her condition. "The textbooks proved of no help to me," he recounted in a medical journal years later, adding that consultations with cardiologist colleagues worldwide also initially proved fruitless. That was until he spoke to a

Norwegian colleague, Anton Jervell, who told him about the patients he had discovered in Oslo with deafness and sudden death. Might Dr. Ward's patient have a long QT interval, as they did?

After speaking with Jervell, Ward went straight to look at the ECG of the little girl. When he measured the QT interval, he found it prolonged. Clearly, he was dealing with a similar condition. Recognizing the importance of family inheritance as described by Jervell and Lange-Nielsen, Ward paid particular attention to this girl's family tree, and he brought her relatives in for testing. It turned out her mother and one brother also had prolonged QT intervals on their ECGs. But her father and another brother's QT intervals were normal. Recalling that each person receives one copy of each gene from each parent, Ward concluded that in this family, where the disease had clearly passed from mother to child, only one altered copy of the gene was needed to cause the disease. This was known as *dominant* inheritance (in contrast with the *recessive* inheritance of Jervell Lange-Nielsen, where two bad copies of the gene are required to cause disease). He also noted the absence of deafness in this girl and her family, further distinguishing the two syndromes. Finally, he highlighted one very important aspect: an adrenaline surge caused by stress was critical to triggering the torsades rhythm.

Professor Ward published his case report in the *Irish Medical Journal,* but it was its highlighting by an anonymous commentator in the pages of another journal—one with much broader circulation—that ensured Dr. Ward's name would be associated with this condition for many decades. That commentary, published in *The Lancet* in 1964, was read by a Genoese pediatrician named Caesaro Romano, who had described a syndrome of fainting, long QT, and ventricular fibrillation. Realizing these were one and the same thing, the medical community gave both physicians credit and Romano-Ward syndrome was born.

The term *long QT syndrome* was not coined until years later, but that phrase has now come to replace the eponymous syndromes from which it grew. This renaming was accelerated by the discovery of the genetic and electrical underpinnings of the condition. In the mid- to late 1990s, the first genes known to be involved in long QT syndrome were identified. Those genes turned out to code for electrical channels in the heart, responsible for moving electrically active molecules like sodium, potassium, and calcium in or out of the cell. And not surprisingly, given that the QT interval occurs when the heart resets to get ready for the next beat, those channels are directly responsible for the electrical resetting of the heart after each muscle contraction.

The major gene responsible for Jervell and Lange-Nielsen syndrome is *KCNQ1*. It codes for a channel that shuttles potassium (K) across the cell membrane. With the identification of this gene, 150 years after the syndrome was first described by Meissner, the reason for the sudden death of that young deaf girl in front of her classmates was finally made clear. So, too, was the curious link between deafness and sudden death. The potassium channel encoded by *KCNQ1* is expressed in only two types of cells: heart cells and the cells of the inner ear. Severe disruption of potassium flow caused by gene mutations in *KCNQ1* can cause sudden death, when a long QT leads to cardiac arrest, but it also causes deafness by disrupting sound detection in the inner ear. And here's the catch: both copies of the gene must be disrupted (one inherited from each parent) to cause deafness. If just one of a person's two *KCNQ1* genes still functions, there is no hearing loss—only the long QT syndrome.

Back in 2014, Jazlene, as her physicians soon learned, was suffering recurrent episodes of torsades de pointes. Each time, the team would come to the bedside, watch for a few seconds, then start CPR. Everyone would watch, holding their breaths. Her pulse would come back. Her fragile body was holding on.

Every time a shock is given or CPR is delivered, however, the heart is stunned and weakened. Jazlene's team moved to provide a more definitive solution. First, Anne Dubin, our chief baby electrician, spoke to Katsuhide Maeda ("Katz" to his friends), one of the surgeons specializing in baby heart surgery. Implantable defibrillators had, by this time, been used in millions of adults, infants, and children around the world, but never in a baby quite this young. Would he be prepared to implant one in Jazlene? Despite the risks, he agreed. Although technological advance has miniaturized many electronics, batteries powerful enough to generate an electrical shock that can reset a human heart remain large—and a defibrillator can only be shrunk as small as the battery. In a newborn baby, there is no room for a defibrillator under the collarbone or under the left arm next to the ribs, where most adults have theirs placed. Room needs to be made in the abdomen by pushing the intestines to one side. Fortunately, the intestines don't mind. The cable that delivers the lifesaving shocks is then tunneled up to the heart from the abdomen. On the day of her birth, Jazlene became, to our knowledge, the youngest baby ever to have a defibrillator implanted.

This alone was not enough. The team also needed to find a way to calm Jazlene's heart. She continued to have runs of torsades de pointes, despite the fact that Anne had maxed out her medication and that the surgeons had also cut the electrical cord to her heart to try to reduce the adrenaline being delivered directly to it. When the stakes are this high and when we don't know the underlying genetic cause of a patient's long QT syndrome, we treat with everything we've got, even though it's possible some of those drugs could work against each other. We give drugs known to block the effect of adrenaline surges (these particularly help with *potassium channel* long QT) as well as drugs that dampen heart muscle activation (these help with *sodium channel* long QT), anesthetics that reduce the brain's fight-or-flight response, and anything else we can think of. In Jazlene's case, we even contacted a local drug company that was developing a new drug not yet approved by the FDA. But their drug was not available in intravenous or liquid formulation, only in pill form, so we couldn't administer it to a newborn.

If only we knew the underlying genetic abnormality, we would be able to target Jazlene's therapy more precisely. But how fast could we get that information? At the time, in early 2014, the standard approach to genetic testing for long QT was to sequence a handful of genes at a cost of around $5,000. The turnaround on that test was twelve weeks. We didn't have weeks; we barely had hours. The team was making and revising decisions every shift, and the clock was ticking.

We wondered if there was another way. We knew that, in partnership with Stephen Kingsmore, an Irish pathologist at Children's Mercy Hospital in Kansas City, Illumina had developed a protocol to sequence a patient's genome in fifty hours. Kingsmore had applied this in the neonatal intensive care unit, where life-or-death decisions often need to be made urgently. If we could persuade Illumina to prioritize a sample from us, and meanwhile speed up our own computational pipeline, perhaps we could get an answer quickly enough to make a difference. I called my colleague Tina Hambuch, who ran the clinical laboratory at Illumina, and who had developed the rapid turnaround protocol. Could she help? The cogs started to turn—they would need to put aside one of their machines entirely; they would need to adjust their sequencing timetable for the week; it would require staff to come in over the weekend. They didn't hesitate. How fast could we get them the blood?

At Stanford, James Priest sprang into action. James was training to be a pediatric cardiologist and, at the time, was a postdoctoral scholar in my laboratory. A talented french horn player, he grew up mostly in Oregon and graduated from Oberlin College & Conservatory, with degrees in both science and music. While pursuing a master's in molecular biology at UC–Berkeley, he cut his genomics teeth in Edward Rubin's lab at Lawrence Berkeley National Laboratory where he worked on chromosome 5 for the Human Genome Project. He then came to Stanford for medical school. His drive to follow his grandfather, a pediatric cardiothoracic surgeon, into medicine and, specifically, into pediatrics came from witnessing, at age seventeen, his youngest sister receive a bone marrow transplant for leukemia. Like many great physician-scientists with musical talent, he brings to his professional life a mix of creativity, nervous energy, and dogged determinism. This potent combination was exactly what was needed right now.

Meanwhile, it was time to talk to the family. Since we would be sequencing this infant's whole genome, genetic counseling was crucial. This was still early days in the use of genome sequencing in medicine. Very few patients had been sequenced, and far fewer newborn babies, so it was really important for the family to have a clear understanding of all the things we might and might not find. Kyla Dunn, our genetic counselor for pediatric heart conditions, was the perfect person for the job. A Yale graduate with a penchant for Japanese hot springs, she majored in neuroscience and worked in biotech, briefly, doing protein engineering before turning her exceptional talent to science journalism. She is the only person I know with a Peabody Award (she also has an Emmy, which she won as an associate producer for an episode of *Frontline*). As well as being represented on PubMed, the database for academic publications (including for a *Nature* paper she coauthored while in biotech), she is on IMDb, the Internet Movie Database. Despite her considerable success as a writer and producer, she yearned to be part of the stories she was telling, so she came to Stanford to train in genetic counseling and joined our team immediately after she graduated as our first pediatric cardiovascular genetic counselor. Kyla headed over to sketch out a detailed family tree and to discuss with Jazlene's parents what we were offering to do. They agreed without hesitation. James oversaw the blood draw, and the sample was on its way to Illumina in San Diego.

Over the next few days, while Jazlene's genome was being sequenced, James worked to optimize our data analysis pipeline for speed. At the time, we used a combination of commercial software from a company appropriately called Real Time Genomics (RTG). The company was based in New Zealand and stood out for its emphasis on industrial software design. They focused on speed and accuracy. We needed both. Once the variants were reported by the RTG program, we used software developed in my lab to parse and prioritize the variants. James started working on making it run faster.

A few short days later, the data arrived in a UPS envelope: a hard drive with Jazlene's genome on it. It was the product of a sequencing run that took just twenty-eight hours. With the Illumina technology at the time, it would barely have been possible to go any faster. Instead, the opportunity to speed up the process came in the alignment of the millions of one-hundred-letter genetic "words" output by the Illumina sequencer to the human reference sequence and in the identification of the genetic variations themselves. Standard approaches, at that time, took twenty-four hours or more. The fast protocol previously used in an ICU setting by Dr. Kingsmore took twenty hours, and he had suggested in a paper that this part of the pipeline could be cut further. Challenge accepted! James halved that to ten hours.

There is one major advantage in analyzing a genome when you have a possible diagnosis in mind. You still need to process each of the millions of hundred-letter words. And generally, you will still catalog all the places where the genome in question varies from the reference sequence. But then, instead of having to parse and prioritize every one of those, you can start with genes already known to cause the disease you are focused on, moving quickly in your hunt for the smoking gun. In Jazlene's case, of course, the most important genes to look at first were the ones known to cause long QT syndrome.

It didn't take James long to find it: Jazlene had a variant in the potassium channel gene, *KCNH2*, associated with long QT syndrome.

Even more convincing, this specific variant had previously been reported to cause long QT in other patients. The gene, otherwise known as *hERG*, has an interesting history. When flies without the gene were exposed to the anesthetic gas ether, they appeared to dance in a way that resembled the go-go dancing common in the '60s at the Whisky à Go-Go bar in Los Angeles. Hence, hERG stands for *human ether à go-go-related gene.*

Although Jazlene's potassium channel gene variant was persuasive as a cause for her long QT, we soon learned that her father also had it, yet he had no obvious signs of long QT. We had to be missing something. Was there another factor, another genetic variant making Jazlene's presentation at such a young age so life-threatening and dramatic?

Combing through the rest of Jazlene's genome, James found another variant that looked like it might contribute significantly. This gene, *RNF207,* was clearly associated with normal variation in the QT interval but had not previously been shown to play a role in long QT syndrome families. However, it looked particularly suspicious as a contributor, partly because Jazlene's actual variant severely disrupted the gene—and because that gene's role includes stabilizing the *KCNH2* potassium channel (produced from the *hERG* gene mentioned above). The variant was also rare enough that it had never been reported before, though we did find it in Jazlene's mother, who did not suffer from long QT syndrome. It seemed Jazlene had inherited two important variants, one from each parent, and it was likely this combination that provoked the severe long QT and torsades de pointes.

James and Kyla took our analysis to the family and to the rest of the medical team looking after ten-day-old Jazlene in the ICU. At the time, Anne was still treating her with every drug available, including a drug that blocked the sodium channel involved in one specific type of long QT. We now knew from her genome, however, that this sodium channel was normal in Jazlene. This particular drug can be toxic if the dose is too high or too low, and dosing in infants is challenging. So here was an opportunity to practice more "precise" medicine. The team was now able to pull back and stop the sodium channel blocker and to focus instead on maximizing other medications that were known to be more effective for the potassium-channel version of long QT.

Over the following months and years, we were able to better understand exactly how Jazlene's genetic variants interacted to cause her condition. One of our collaborators at the University of California at Davis, Nipavan "Nip" Chiamvimonvat, helped enormously with a series of electrical studies. In her lab, Nip studies single channels in single cells. Using the information from Jazlene's genome, Nip made a series of different cells to study. Some were cells with normal potassium channels, while others had the mutated *KCNH2* channels that were contributing to Jazlene's long QT syndrome. Through a careful series of experiments, Nip showed that a normal *RNF207* protein could make up for the reduction in potassium current from the mutated *KCNH2* channel. This was exactly the situation in Jazlene's dad. He had a normal *RNF207* gene, but a mutated *KCNH2*. This experiment might explain why he was living a normal life with no outward signs of long QT. When, however, Nip looked at the cells that had both mutations, just like Jazlene, she found that the mutated *RNF207* was unable to make up for reduced potassium current, and an *extremely* severe long QT occurred. This was exactly what we saw in Jazlene's heart. In the game of genetic roulette, this little one had received two variants affecting the electrical resetting of her heart, each of which made the other worse.

Twelve months later, we were excited to see Jazlene now toddling around, playing, and giggling in a local park. Her defibrillator was still watching her, protecting her, but fortunately, she hadn't needed a shock from it in months. Playing on the roundabouts and swings, she had no idea how much technology and care had been required to keep her alive in the earliest weeks of her life. Her genome had revealed its secrets but rather than determine her future, that knowledge could now provide her the power to control her destiny. What breakthroughs might come to redefine our approach to long QT syndrome in her lifetime? Swirling around that day, with the wind in her hair, that was something that could wait for another day.

# 13

# How Many Genomes Are You?

*Oh, ye canny shove yer grannie aff a bus*
*Oh, ye canny shove yer grannie aff a bus*
*Oh, ye canny shove yer grannie*
*Cuz she's yer mammie's mammie*
*Oh, ye canny shove yer grannie aff a bus.*

—TRADITIONAL GLASWEGIAN SONG,
SUNG TO THE TUNE OF "SHE'LL BE COMING ROUND THE MOUNTAIN"

*We are mosaics.*
*Pieces of light, love, history, stars,*
*glued together with magic and music and words.*

—ANITA KRIZZAN

There is an expression from my homeland in Glasgow, Scotland, about number 3 buses. It comes from the observation that, when standing at a bus stop, no number 3 buses come for what seems like an eternity, then a whole bunch come at once. Although none of us can explain it, rare cases in medicine also seem to bunch together. Long QT syndrome is rare to begin with (about 1:2,000 in the population) and neonatal long QT with heart block and torsades de pointes, well, that is very rare indeed. Yet hard as it was to believe, within the space of just a few weeks, another baby presented to us with an almost identical story. Because of a low fetal heart rate, her mother was transferred to Stanford, and baby Astrea was born early by cesarean delivery, suffering multiple cardiac arrests early in life. She also received very early surgical implantation of a defibrillator. Our friends at Illumina worked their magic, and within a few days, we had her genome on a hard drive.

But baby Astrea's case was not going to be solved so easily. We ran our optimized computer software, and as was the case for baby Jazlene, there was one particularly suspicious variant in a gene known to cause long QT syndrome. In some ways, this variant was initially more convincing than Jazlene's variant. It was found in the sodium channel gene *SCN5A,* at the exact location where another variant had previously been reported to cause not just long QT but a neonatal presentation of long QT with torsades de pointes. In that same study, the team had gone as far as to measure the exact magnitude of the electrical current through the altered sodium channels in cells in the lab, which they found to be highly abnormal. Surely, we had found the smoking gun?

Not so fast.

When analyzing a genome, we usually use several different computer approaches to identify variants so that we have confidence in the variants we actually declare. Usually, those approaches agree. Yet in this case, only one of the three approaches included the smoking gun gene variant in its final list. That was puzzling. Plus, one out of three is not sufficient when the stakes are this high. We want to be certain. So before even mentioning the finding to the intensive care unit team, we set about doing old-school validation with Sanger sequencing to be sure the *SCN5A* variant was really there.

It was the weekend (of course), and James Priest, future pediatric cardiologist, was busy working in my lab. Given how convincing this variant was, we were not really expecting anything other than a clear confirmation from the old-school method. We were planning to share the good news with the ICU team later that day so they could begin personalizing Astrea's care by focusing their therapy on blocking the sodium channel. But that wasn't what happened. Not by a long shot.

The Sanger reaction had worked well, and the readout of the gene's As, Ts, Gs, and Cs, a series of colored peaks like psychedelic mountain range, was clean. Too clean in fact: it didn't show the variant. We were dumbfounded. How could that be? So James repeated the Sanger confirmation process, this time isolating DNA from Astrea's saliva as well as from her blood, but he found the same thing. Which is to say, he found nothing.

To investigate further, we looked at the raw data from the genome sequencing that had covered each position twenty-five times. We found that of the twenty-five predictions for the letter at that position, twenty were the same letter as the reference, while five contained the variant letter. So now we understood why the different computer programs disagreed. The programs, at that time, were designed to look for only two possibilities: 1) if the two copies of the gene received from Mom and Dad are identical at that position, then all DNA reads should show the same letter at that position; 2) if the copies from Mom and Dad have a different letter at that position, about half of the reads should show Mom's letter and half should show Dad's letter. Of course, random chance means it is rarely exactly 50 percent. But for one letter to be represented by just 20 percent of reads? Sure, if you flip a coin many times, you won't always get an even split, but what are the chances that you get only five heads out of twenty-five flips by chance alone? It turns out the chances are not very high—about one in a thousand. Something was wrong.

Our first thought was that this could be a problem with the alignment of the reads to the reference. This instinct came from our earliest experience with Richie Quake's genome (in chapter 3), where our first answer turned out to be a mapping error due to a pseudogene (an area of the genome that looks like a real gene but isn't). Our mantra became "If the signal isn't clean, it must be mapping." And this is especially true for electrical channels like *SCN5A* that have a lot of repeated sequences, many closely related gene family members, and pseudogenes. They are hard genes to map accurately. James took on the job of painstakingly mapping the reads in question by hand to all the areas in the genome that were similar enough to have caused confusion. His conclusion was clear: the reads were mapped to the right genome segment.

I called Rich Chen, the chief scientific officer of Personalis, our new sequencing start-up company. Because it was Saturday, Rich was at his daughter's soccer game. As I relayed the story and our confusion over what was going on, I could hear his daughter's team dribble and tackle and cross and shoot in the background. He offered to pull Personalis staff in to run deep sequencing of baby Astrea and her parents. The deeper sequencing would give us more than just twenty-five glimpses of the DNA letter at this location and hopefully clarify that we weren't on a wild-goose chase for a variant present in just five errant reads. Sequencing Astrea's parents would also help us to determine whether whatever this was had been inherited or whether it arose for the first time in Astrea.

When John West, the CEO of Personalis, called me that weekend to talk through the puzzle, our minds turned to a rare potential explanation. The phenomenon was well recognized in certain other genetic diseases, and especially in some affecting the brain, but what we were considering had not been observed for a genetic cardiovascular disease before.

Thanks to fast couriers and the folks at Personalis prioritizing the case, we got back the data from the deep sequencing of all three family members in just a few days. James checked the location of the variant. This time, that spot was covered 210 times; 17 matched the variant, while the rest matched the reference. In short, the same skewed signal was there. It was high-quality data, but this time, the percentage (8 percent) was even further away from the 50 percent expected if one of the two copies of Astrea's *SCN5A* gene was mutated. The chances that this was a normally inherited variant were vanishingly small, a finding underlined by the fact that the variant was not present in either of her parents.

Our alternative theory was now firmly at the top of our list: Was it possible that Astrea had more than one genome? Could she be a rare example of a genetic mosaic, with the disease-causing *SNC5A* variant present only in certain cells of her body? And—what seemed even more of a stretch to us at the time—could such a small minority of affected cells cause such severe disease? Over the next few weeks, we set out to try to find answers.

The scientific term for having more than one genome is *somatic mosaicism* (*somatic* comes from the Greek word for "body"). Mosaicism happens because, as cells divide over and over in development or in life, their oft-copied DNA accumulates mutations. If a genetic mutation occurs early in development, after the egg and sperm come together to make an embryo, the cell with the mutation then gives rise to a whole group of cells with the new genome. Eventually, these cells will be found in a variety of tissues throughout the body.

In some people, there are actually visual indicators of mosaicism: patches of skin with different pigmentation or eyes of different colors. Mosaicism can also become visible when the altered cells start to grow out of control.

Leslie Biesecker, geneticist at the National Institutes of Health, was one of the pioneers of our understanding of mosaicism. He reported on a group of patients with multiple genomes and a condition called *Proteus syndrome.* These patients present disfigured by progressive, irregular growth of many tissues—most commonly bone, connective tissue, and fat. It appears quite likely that Joseph Merrick, known as the "Elephant Man," suffered from Proteus syndrome. In 2011, Biesecker reported the striking finding that in a group of twenty-nine patients with this condition, twenty-six had the very same somatic mutation in the gene *AKT1,* but only in the affected parts of their body. Studies of cells from these patients showed that growth circuits were turned on only in the mutated cells, explaining both the growth and its pattern.

Another well-recognized example is neurofibromatosis—a disease that affects the brain, spinal cord, nerves, and skin. In neurofibromatosis, benign tumors grow along the nerves and can be felt under the skin. In the most common version of the disease, these tumors can occur anywhere, but in the rarer, mosaic form, the changes are *segmental,* which is to say that some parts of the body are affected and some are spared. When biopsies of these different areas are performed, differences in the genetic makeup of each area can be detected.

Mosaicism has also been characterized in the brain. Christopher Walsh, a neurologist at Harvard Medical School and Boston Children's Hospital, reported that somatic mutations cause a syndrome that leads to enlargement of just one half of the brain.

Your bone marrow, too, the red gelatinous substance responsible for forming blood cells, can be mosaic. Its stem cells generate more than one hundred billion new blood cells every day, giving those rapidly dividing stem cells a lot of opportunities to accumulate mutations. Indeed, mutations in blood stem cells appear to predict early death. In one study that sequenced blood stem cells from thousands of individuals, the presence of new genetic variation in a group of blood stem cell–related genes was associated with a 40 percent increase in the chance of death. Although you might think most of those deaths were explained by blood cancer, in fact, most were explained by heart attacks and strokes. In a follow-up study, investigators noted a two- to fourfold increase in the risk of heart disease or heart attack in patients with demonstrated new and expanding mosaic blood cell lines.

Mosaicism is not the only way to have more than one genome, however. Whereas in mosaicism, a person's own cells start to differ by accumulating mutations, *chimerism* describes a situation where a person has cells with genomes derived from completely different people. The chimera was the hybrid animal from Greek mythology with a lion at the front, a snake at the back, and a goat in the middle. The chimera also somehow managed to breathe fire. Despite the monstrous origins for the word, chimerism in some forms is very common. For example, all women who have previously been pregnant are chimeras because they carry a small number of cells from each of their children around in their bloodstreams. Also, any patient who has received an organ transplant is a chimera, because the transplanted organ has a different genome. It can also happen that nonidentical twins who share a placenta absorb some blood cells from each other because of mixing of blood in the common placenta.

Chimerism can lead to some strange occurrences, such as one reported by science journalist Carl Zimmer. He described a woman who, as part of a workup for a kidney transplant where she and her immediate relatives had blood tests to determine who would be the best match, learned that she was not the genetic mother of two of her three biological children. It turned out that she herself was a fusion of two genomes: one genome gave rise to her blood and to some of the eggs in her ovaries; other eggs carried a completely different genome. Explanations for how such chimerism arises are debated. They include the possibility of one person arising from the fusion of two embryos early in their development or from one egg fertilized by two different sperm.

James Lupski, the Baylor College of Medicine geneticist who published his own whole genome in late 2010, predicts an increasing medical role for detecting our hidden genomes. He believes the future will be one where we find mosaicism much more often by paying attention to the tissue of origin of the DNA in question. "Eventually, genome analysis of all surgically excised abnormal tissue . . . not just cancer, might be considered germane for genome analysis to detect mosaicism," he wrote. Like so much of our dogma, the elegant simplicity of "every cell contains the same genome" turns out, in many cases, simply not to be true.

. . .

In the case of baby Astrea, we now had a challenge: How could we prove our suspicion that she was a mosaic and that her long QT syndrome was due to an *SCN5A* variant present only in some of the cells in her heart? A biopsy of a neonate's heart was not advisable. Yet in embryonic development, blood cells are derived from the same original tissue as heart cells—and blood cells are easy to come by. If we had a way of doing individual genetic tests on single white blood cells, we could establish definitively that these cells in Astrea had differing genomes.

Fortunately, someone we knew rather well had developed an approach to allow us to investigate the genetics of single cells. Steve Quake had invented a way to isolate single cells in microchannels and perform tests on them and had even founded a company, Fluidigm, based on the technology. The technology had not, to our knowledge, ever been used to diagnose a patient before—it was used in research labs to help scientists understand the biology of single cells.

James connected with Chuck Gawad, a researcher in Steve's lab. After some troubleshooting, they were able to run Astrea's white blood cells through the system, a series of microchannels through which you can flow very small volumes of liquid. More important yet is the use of valves to isolate not just single cells but also the minute volumes of liquid required to isolate DNA and amplify it into large enough quantities to allow us to run a genetic test on it. After a few weeks, they successfully isolated good-quality DNA from Astrea's individual cells. Now it just needed to be tested.

When the analysis came back, it clearly showed two distinct populations of cells. Our intuition had been right. Eight percent of the cells isolated from one sample of Astrea's blood harbored the variant in the *SCN5A* gene that we believed was causing her condition. The other 92 percent did not. What had been impossible to confirm with certainty from a pool of DNA derived from many cells was now clear from our testing of individual intact ones. We were delighted to share the results with the team looking after Astrea in the intensive care unit.

Understanding that the cause of her long QT syndrome was a faulty sodium channel would allow the team to use drugs targeted to the exact mechanism of disease. For her, unlike baby Jazlene, a drug that blocked the overactive sodium channel was the precision medicine she needed, worth navigating the challenges of dosing it in infants. We were excited by the possibility of her long-term health with fewer side effects.

But it was not to be. A few months later, Astrea took a turn for the worse. Her defibrillator fired to save her from a dangerous heart rhythm, and she was called back to the hospital. There, tests showed her heart had weakened and enlarged, a condition that can sometimes result from mutations in this gene. Baby Astrea was now dying from heart failure, and only a heart transplant would save her. However, the team were uncertain whether she would even survive the wait for a donor; she needed a bridge to transplant with an artificial heart assist device. These devices are pumps that can live inside or outside the body. In adults, internal devices allow patients to leave the hospital, climb mountains, and even ski. For babies, the only option is a group of devices coming in different sizes called the Berlin Heart ventricular assist device—Berlin Heart or even Berlin, for short, after the company that manufactures it. These pumps, about the size of an adult fist, live outside the body and draw blood from one or both sides of the heart and pump it continuously to the body. Astrea's device was implanted a few days later on a Saturday afternoon.

I remember the day well. I headed into the hospital around lunchtime to check on her progress and to help collect the tiny piece of her heart tissue that got removed when they inserted the pump. My student divided up that tiny bit of heart and carefully stored the pieces to preserve DNA, RNA, and protein. Then we waited anxiously to see how her little body would respond to the pump. Thankfully, the news was good. She stabilized and started the long, uncertain wait for a heart donor.

Meanwhile, in the lab, the most important thing we could do for her was to try to confirm our explanation for her disease: Would we see the same mosaicism in her individual heart cells that we had seen in her blood? Would the percentage of altered cells be the same? With a heart sample, we could not only look at DNA but actually look at the RNA message used to build Astrea's heart's sodium channels. This is important because sometimes cells can police their own mutant messages and shut them down at source. The Personalis team stepped up again. They looked at the RNA in two tiny pieces of her heart. There was no easy way to isolate single cells from the heart, so we were looking at all the RNA in the tissue from all cells—the message is what is actually used to build the sodium channel, so we were interested to see what percentage of RNA was from mutant channel genes versus normal channel genes. In one piece, 5 percent of RNA was variant; in the other, 12 percent was variant. This confirmed three really important things: 1) mosaicism was definitively present in Astrea's heart, 2) it was reflected in the RNA message, not just the DNA code, and 3) the proportion of mosaicism varied within two pieces of tissue from a similar part of her heart, suggesting that the cells with the alternate genome were not evenly distributed throughout the tissue.

One big question remained: Could such a small percentage of mutant cells actually cause severe long QT?

The first part of this question was all about the gene variant itself. Could we establish that it disrupted the sodium current in a dramatic way? We knew already from a prior publication that another mutation at this exact position disrupted channel function, but the only way to be sure that Astrea's mutation could similarly disrupt the channel was to make her version in some cells and measure the sodium current in those cells. We got in touch with colleagues at Gilead Sciences who had a lot of experience measuring sodium currents in mutant sodium channels. In fact, this was the same group we contacted to ask about the new sodium channel drug they were developing. Would they be able to help? Like everyone else we called on behalf of these little babies, they didn't hesitate. We sent the *SCN5A* variant details, and they set to work. When the team measured the sodium current from Astrea's channel, they found it resulted in far more sodium current than is normal—more, in fact, than any variant they had ever seen.

So clearly, the variant was important in those cells in which it appeared, but we still had no handle on whether such a small number of affected cells, just 5–12 percent, could produce life-threatening long QT syndrome in a whole heart. Wouldn't the more-numerous healthy heart cells prevent it? Perhaps, we decided, a computer model of a heart could help.

We knew it was possible to model single cardiac cells, including the flow of electricity through channels from genes like *SCN5A*, and we knew a small number of groups had begun to stitch these single-cell models together to generate models of a whole heart's electrical behavior. But to our knowledge, no one had ever modeled a baby's heart, certainly not with a view toward using it to inform patient care. And no one would have had any reason before now to consider modeling cardiac mosaicism in a heart with long QT. Wherever we were, we were in uncharted territory. It was once again time to call a friend.

One of the world leaders in this form of cardiac electrical modeling is Natalia Trayanova at Johns Hopkins University in Baltimore. Natalia was born in Bulgaria and was interested in rockets and propulsion from a young age. After one trip to the United States, her dad brought her back a book on bioelectrics (electrical activity in heart, muscle, and nerve cells), and it sparked a lifelong interest in the electrical activation of the heart. She later attended Duke University for a fellowship with the author of that book, and her passion and drive for understanding the heart grew and grew. There, she was named Murray B. Sachs Professor of Biomedical Engineering, the first endowed chair for a female faculty member at the Whiting School of Engineering at Johns Hopkins. In 2019, the Heart Rhythm Society awarded her its Distinguished Scientist Award. Quite a journey from behind the iron curtain to center stage in the global science of cardiovascular bioengineering.

I posed this question to Natalia: Did she think her computer models of the heart could be tweaked to model a mosaic heart? A mosaic baby's heart? And if so, would her team be willing to try? Once again, without hesitation, she responded yes to all three.

And so, over the course of the next few weeks and months, Natalia's team, led by bioengineer Patrick Boyle, adjusted their model to reflect what we knew of baby Astrea's heart: the proportion of mosaic cells in the heart; the alteration in sodium channel current reflected by our cell experiments. They had the great idea of modeling different patterns of mosaic cells within the heart, one model with the cells clumped together, one with them scattered like salt and pepper. These computer models are 90 percent preparation and planning and 10 percent sending the computer off to do its thing. Finally, after months of planning, they were ready to run the models. The computer headed off only to come back with . . . nothing.

The computer-modeled heart appeared normal. No long QT. No low heart rate. No heart block. No life-threatening torsades. Back to the drawing board. Did this truly mean that a small group of mutant cells was not capable of causing such a severe form of long QT? It seemed, by now, so improbable that this variant would not be the cause. Perhaps some aspect of the model was not quite right yet?

It then occurred to all of us that the heart model was entirely made up of muscle cells. Yet in a real heart, there are also those specialized electrical cables that run through the heart, conducting electricity. The cells that make up these cables are the Purkinje cells. Perhaps adding this

Purkinje system to the model would better reflect the speed and pattern of overall electrical activation? And perhaps with this new property, the model would better mimic Astrea's heart. The team redoubled their efforts and, in an extraordinary advance, built in the branching network of electrical cables to their heart model in record time. They sent the computer off again. Suddenly, it was clear: the computer heart was showing not just long QT but also heart block and occasional broadening of the electrical activation, exactly the way Astrea's heart had when she was first diagnosed inside her mother's womb! To watch the video of the 3D heart model, with a rainbow wave of electrical activation washing over the virtual tissue, was truly astonishing—mathematical and aesthetic beauty in perfect harmony—as mesmerizing as the first heart I saw beat outside of the body when I was a medical student in my physiology class.

Finally, after these many months, I was ready to believe that this extraordinary team really had cracked the case of Astrea's long QT syndrome. From Stanford doctors, surgeons, scientists, and genetic counselors, to industry partners at Illumina, Personalis, and Gilead, to bioengineers at Johns Hopkins, to James with his perseverance and dogged determinism, everyone contributed something unique that only they could. Astrea became one of the youngest babies ever to have a defibrillator implanted. Her genome was sequenced and analyzed as fast as humanly possible. She was the first person to have a single-cell genetics test from her blood, and the first baby to have a personalized computer model of her heart built as part of her medical care. We published the paper describing these findings in the *Proceedings of the National Academy of Sciences* (one of the reviewers was mosaicism pioneer Les Biesecker, whose constructive feedback greatly improved the final paper).

In the end, Astrea only needed the Berlin heart pump for a few weeks. Five weeks after being listed for transplant, a donor heart became available. Thanks to this new heart, complete with its boring single genome, she is now both a mosaic and a chimera. Another first for this special little one. I caught up with Astrea's mom in late 2019. Astrea had just turned six years old. She was enjoying kindergarten and loving gymnastics and ballet. As her mom put it, after all she's been through, "She's now one happy big girl."

# 14

# Shake, Rattle, and Roll

*Treat me nice, treat me good*
*Treat me like you really should*
*'Cause I'm not made of wood*
*And I don't have a wooden heart*

—BERTHOLD KAEMPFERT / KAY TWOMEY / BEN WEISMAN / FRED WISE
(AS PERFORMED BY ELVIS PRESLEY)

*Of the many things hidden from the knowledge of man,*
*nothing is more unintelligible than the human heart.*

—HOMER, *The Odyssey*

Hey, check this!" said my cardiology fellow Rick Dewey, who had just pulled up the echocardiogram of our next patient. The rest of the team stopped what they were doing and rolled their chairs over to huddle around the computer screen. Immediately, everyone was fixated. Sure, the heart walls looked good. The chamber size was not too big or small. The pumping strength of the heart was good. The valves opened and closed just right. But inside the cavity of the main pumping chamber, the left ventricle, was something that none of us could explain. As we stared, transfixed, we saw two little muscle balls tethered to the heart wall by slender stalks. They were dancing around, bobbing, surfing the wave of blood that gushed past on its way out of the heart. They were strange and improbable—impostors—they simply shouldn't have been there.

Even for a clinic specializing in rare cardiac disease, and a medical team accustomed to patients who are sometimes quite literally unique in the world, it was not common to see tumors in the heart. It was uncommon to see just one, never mind two. And these, frankly, had us concerned; they were flicking back and forth at the exit door of the heart, from where the road leads directly to the brain. I had never seen anything like it.

Unlike some of our other organs, you see, the heart is made up mostly of cells that don't divide. If you damage your liver or your skin, new cells will form to repair the damage. Your heart simply can't do that. That's why a heart attack, when a blockage in a coronary artery deprives heart cells of blood and oxygen long enough that they die, can be such a devastating event. Those cells don't come back, and new ones can't be generated to replace them. Heart cells are, in technical language, *terminally differentiated.* Since terminally differentiated cells don't divide, that also means they generally don't form tumors. (Tumors arise when cells divide, and new mutations form that allow unrestricted growth.) I stared longer at the screen. "How is that even possible?" I found myself thinking out loud.

Well, Ricky, our patient, was an impossible sort of guy. This was his first visit with us, at age twenty-one, and he was transitioning to us from the children's cardiology clinic. He was quiet and shy, but greeted us with a huge smile and a firm handshake. He was also tall and dressed all in black, with long, dark hair pulled into a ponytail. He had a lithe frame that sported some tattoos and a few body piercings. His pediatrician had described his style as goth, but in reality, Ricky was entirely his own genre; his personality was as distinctive as his heart.

And what about that heart? The first thing to know is that Ricky's tumors were not malignant. Malignant tumors, what we usually mean when we talk about cancer, can spread or *metastasize.* It is this invasion of normal tissue that makes cancers so lethal. In contrast, Ricky's tumors are so-called benign. Although, they are really only benign by comparison. True, they don't invade and destroy normal tissue, and they don't spread around the body, but they are capable of causing major problems. They do this by compressing the surrounding heart tissue. In fact, when benign tumors do not have surrounding tissue limiting their growth, they can grow to be very large indeed. (Ovarian tumors are not uncommonly tens of pounds in weight; the largest ever removed was over three hundred pounds.) Most, however, tend to arise in a tight spot where there is little room to grow. Cardiac tumors are usually a specific type called a *myxoma,* and they have characteristics somewhere in between these two examples. The pressurized blood inside the heart compresses them and makes them dance around but doesn't limit their growth quite as much as solid tissue would, since they are essentially dangling inside a pool of liquid.

Ricky had first presented to his local hospital when he was seven years old. He had complained of being tired. His mother also noticed something unusual. "He would wake up, go straight to bed," she recalls. "No energy, no appetite. And then I noticed the palpitations: boom, boom, boom, boom!" As she described these palpitations, she showed me the pounding she had seen in her little boy's neck all those years before. To Ricky, those palpitations felt like he was "running or something, and I was just sitting there."

During one episode when Ricky had a cold, his mom took him to the local doctor. On that day, his regular doctor was out, and the doctor filling in felt she should do a more thorough exam than normal because she didn't know Ricky very well. To her surprise, she heard a loud and unusual whooshing sound in his heart. That wasn't normal. Grateful that she had decided to do a more detailed examination, she urgently referred him to Stanford's Lucile Packard Children's Hospital, where he met my colleague Daniel "Dan" Bernstein. Dan is a tall, bearded, warm, and gracious pediatric cardiologist, as well as a brilliant scientist who led the pediatric cardiology division at Stanford for many years before diving deeper into science as the stem cell revolution was taking off. Dan takes care of kids with heart failure and is an expert both on the mechanical workings of the heart and on heart transplantation in children. When I spoke to him about Ricky, Dan vividly recalled their first meeting a decade earlier. He remembered Ricky's big smile and also the large tumor at the top of Ricky's heart, in the right atrium, and how it flopped back and forth, obstructing blood flow across the tricuspid valve. Surgery wasn't a hard decision to make.

Everything went well, although the surgeon needed to use a patch to repair the area of the heart from where the tumor had been removed. Seven-year-old Ricky was out of the hospital quickly, discharged in time for the Super Bowl. He recovered well and soon returned to school to finish second grade. The family had no way of knowing at that point if new tumors would appear. Nobody, in fact, seemed to know why his heart had grown the tumor to begin with. But doctors and family alike fervently hoped it would not return.

Unfortunately, that hope was in vain. All went well for a few years. Then, at age ten, Ricky noticed a swelling on one of his testicles, and it turned out to be a testicular tumor. How could that be? To have two rare tumors? The surgeon's knife was once again quickly applied to remove the tumor, fortunately without complication. But now, at a very young age, Ricky had already been afflicted by two rare tumors. They had to be related. Ricky was referred to cancer specialist James "Jim" Ford, who had started one of the first cancer genetics clinics in the country at Stanford. Jim is the personification of the phrase *larger than life.* His laugh is loud and infectious, and when you hear it from across the room, you know that is the conversation you want to be in. Jim knew that not many cancer syndromes include both cardiac tumors and testicular tumors, and when a closer look at Ricky's skin revealed areas of darker pigmentation, a unifying diagnosis seemed likely: Carney complex.

. . .

Carney complex was first described in the 1980s by J. Aidan Carney, a pathologist from Mayo County in Ireland. He attended medical school at University College Dublin, where he studied pathology before moving to the Mayo Clinic in Minnesota to complete a Ph.D. His Ph.D. focused on myosin, the molecular motor of the heart, in an experimental model of cardiac hypertrophy. After his Ph.D., in the early 1960s, he joined the Mayo Clinic pathology department for residency. He had a keen ability to recognize rare patterns in patients, and his career was characterized by repeatedly chasing such associations, sometimes over decades, to describe new diseases. And so it was, in 1981, that he noticed in four patients unusual pigmented areas of skin. These patients all shared a condition caused by too much of the hormone cortisol (a stress hormone produced by the adrenal gland), but it was not entirely clear how these two findings were connected. Carney nevertheless described this combination in a paper and named this new disorder primary pigmented nodular adrenocortical disease (PPNAD). He also observed that the condition seemed to be running in families. In one family, it affected two siblings, one of whom had a stroke in his thirties. Yet a third sibling, who had died at age four, seemed to have a different rare condition entirely: cardiac tumors. Carney thought this was odd. The detective in him set to work on connecting the dots.

Carney believed the skin changes, adrenal tumors, and cardiac tumors in this family must be connected (earlier, we discussed the idea of Occam's razor, the concept that, when confronted with seemingly disparate facts, we should prefer a single simple explanation to several unrelated explanations). He combed the Mayo Clinic patient files and the literature for the co-occurrence of these conditions. In the modern day, this is about as easy as pulling up a Google search bar—but at a time when all medical notes were on paper, this was painstaking work. He had to first identify all the patients with one condition, then pull their charts by hand and go through each page looking for evidence of the other condition. And vice versa. Despite years of working tirelessly through late nights and early mornings, he came up short. But before abandoning the search, he had one last play. In a "desperate effort" to complete what had been a "time-consuming and psychically draining" experience, as he said in an interview, he thought he should review directly for himself the actual microscope slides of the adrenal glands from the cases of patients who had died from cardiac myxoma. The clinical records had not noted any abnormalities, so his hopes were not high, and it was clearly with some resignation that he began this final lap. And then, he saw it. Absent from the clinical record but screaming at him from the microscope slides themselves: one of the patients who died from a cardiac tumor also had adrenal tumors. "The formal clinical record was a revelation. I went through it with mounting excitement and almost disbelief," he said. The medical resident had noted deeply pigmented moles covering most of the patient's body, just as Carney had observed in his patients. "It seemed almost impossible—statistically—that all the patient's conditions . . . could be encountered together," he said, "and the events be unconnected. They must all be related; they must constitute a syndrome." A close shave with Occam's razor, and Carney complex was born.

Despite this detailed clinical description, the molecular nature of Carney complex remained unclear for many years. The disease was clearly inherited, and studies had identified certain large areas of the genome that were suspicious in families, but to narrow it down to the causative gene would require a geneticist trained in family-based studies. One such researcher was Constantine Stratakis, a Greek physician-scientist who grew up in Athens. Stratakis knew he wanted to be a biochemist and geneticist when he was fifteen years old. Attending one of the most prestigious high schools in Athens, and influenced by his uncle, a biologist and one of the founding professors of the University of Crete, his initial interest was in science, not medicine. But when his brother was diagnosed with a pituitary tumor when Stratakis was in his teens, he realized that if he trained in genetics and endocrinology, he could perhaps work out the basis of his brother's tumor. With his uncle's support, he set out to study medicine so he could research the genetic basis for endocrine disease. Remarkably, he is still doing that to this day. The young Stratakis had learned French in Athens, and so, after completing his medical degree in Greece, he continued his studies in France. After that, because half of his mother's family were Greek American and based in D.C., he moved to the United States, training in genetics at Georgetown and finally secured a position in the tenure track at the National Institute for Child Health and Human Development. There, he was on the lookout for a disease he could really sink his teeth into. Some endocrine conditions were beginning to have their genetic bases teased out. By the early 1990s, he realized that, with recent advances in genetic mapping, there was a good chance he could find the genetic cause of Carney complex. Still, he needed Aidan Carney's help and blessing. So in 1994, he nervously wrote Dr. Carney a letter, explaining his interest in the disease, expressing confidence that he could find the gene, and asking for a collaboration. He received no response. This was devastating to the young scientist, but he didn't give up. A year later, he finally managed to connect by phone with the senior clinician, and in February 1995, he took a trip to the Mayo Clinic to meet Dr. Carney in person.

Minnesota in February is not for the fainthearted. During Stratakis's visit, the high in Rochester was 1 degree Fahrenheit, the low minus 11 degrees. "It was good for work," he told me. "We didn't spend a single minute outside!" Stratakis went through the records of all the patients Dr. Carney had seen and started drawing out the family trees. In the 1990s, most gene-finding studies were completed using a linkage mapping approach, where markers spaced along the genome were measured in each affected family member's DNA, narrowing as much as possible the common area responsible for the disease. The DNA-cutting enzymes ("restriction" enzymes) that made this mapping possible were discovered in the 1970s. The technique, known as *genome-wide map-*

*ping,* was pioneered by David Botstein, a geneticist hailing from the Bronx, New York, whose biography reads like a highlight reel of the formative decades of genetics. Transiting through Harvard, MIT, Stanford, Genentech, and Princeton, and currently chief scientific officer of Google's secretive antiaging company, Calico, Botstein's description of the approach, published in 1980, was one of the most influential papers in genetics in the latter part of the twentieth century. If you were a researcher studying a family with many distantly related family members who all shared a family disease, then the location of the shared disease-causing piece of the genome could be narrowed considerably using these maps. If the size of the family being studied was large enough, with enough affected members scattered across enough branches of the family tree, the region of the genome causing the disease could be narrowed to one containing just a small number of genes. Those genes could then be sequenced in each affected family member using Sanger sequencing, to look for a specific shared mutation that might cause the disease. It was a lot of work, but it was simple, and it was brilliant.

Using the family trees for Carney's patients, Stratakis first carried out a simulation analysis—to give him an idea of the likelihood of narrowing the part of the genome responsible to a small enough area to identify the causative gene. Usually, a score of three or four on this test is good. When Stratakis calculated this score for the Carney families, it came back at almost eight. So, now all he had to do was collect DNA from the families and create his linkage map. Between them, he and Carney divided the country up and started driving. They would go to churches and community centers to collect blood samples from family members of known patients or show up at people's houses at Thanksgiving when several family members were in town. Although the research was funded by the NIH, the collection of samples was unfunded. So the gas money came from the investigators' own pockets. "Even the speeding tickets were all my own!" Stratakis told me. All told, it took about a year. After many thousands of miles, with all the DNA in hand, it was finally time to build the genetic map.

Stratakis's mapping first identified a suspicious area of the genome on chromosome 2, but that turned out not to contain a likely gene. However, a second region, identified on chromosome 17, provided the eureka moment. In 2000, Stratakis and Carney reported the gene *PRKAR1A* as the cause of Carney complex. Years of elegant studies by Stratakis and others have since identified the downstream effects of disabling *PRKAR1A,* something that leads to an increase in the function of protein kinase A, a principal actor in the stimulation of tumor growth.

It appeared, then, that Ricky had Carney complex. Jim Ford's team moved quickly to sequence the *PRKAR1A* gene, to try to confirm the diagnosis. The tests available in those days were based on Sanger sequencing, in which the protein-coding sections of the gene are amplified in several different parts, those parts are sequenced, and then each part is compared to a reference. This test was available from a clinical genetic testing lab, but, as for many genetic tests at that time, Ricky's U.S. insurance company was reluctant to pay, so the family had to decide whether they could pull together the thousands of dollars it would cost to sequence one gene. His family raised funds for the test, knowing there was a chance it would not find anything. Jim's team sent off Ricky's blood sample and waited. When the results came back, nothing unusual was seen in Ricky's gene. No mutations. Maybe he had a different form of Carney complex, caused by a still-unidentified gene? Maybe it was another condition altogether? It was unclear.

Ricky did well until the age of thirteen, when a tumor was found in his pituitary gland. The pituitary is like the body's central command center for hormones. It's a small, pea-size gland in the brain that sits just under a part of the brain called the *hypothalamus,* which delivers instructions to stimulate the pituitary. Tumors in the pituitary are usually benign, which is to say that, like the tumors that grew in Ricky's heart, they are slow growing and cause problems more by compression than by invasion. Except unlike in the heart, there isn't a lot of room to grow deep inside the brain. In fact, the optic nerves from the eyes run right past that spot, so enlargement of the pituitary gland can cause visual field problems. Aside from compression, pituitary tumors can also release too much hormone, meaning the patient will suffer from the effects of whatever hormone system is in overdrive.

At the time, thankfully, the pituitary tumor wasn't causing too many problems, so his Stanford endocrinologist, Laurence Katznelson, decided to watch and wait. Three more years went by, and Ricky grew like a weed. Unfortunately, new tumors in his heart started growing as well, and at sixteen years old, he returned to Stanford where it seemed like there was no alternative but to open him up again and remove them. At sixteen years of age, only nine years after his first surgery, it was time for his second open heart surgery.

Ricky's operation went smoothly. The tumors were removed, and once again, his heart was clear of foreign invaders. But this time, it was only to last two years. At age eighteen, he had a recurrence of the cardiac tumors. This time, he also suffered from ventricular tachycardia, that dangerous heart rhythm that can foreshadow sudden death. There seemed to be no alternative but to undergo a third open heart surgery.

Once more, he sailed through surgery and recovered well. Except now, his pituitary was starting to misbehave. Ricky's vision was fine, his optic nerve was not compressed, but the proliferating cells were sending two hormone systems into overdrive. The first was growth hormone and the second was the stress hormone, cortisol, released from the adrenal gland on the pituitary's say-so (the same stress hormone that was overactive in the patients Carney first described). Growth hormone has different effects on growth depending on whether you are still growing or have stopped growing. During the growth spurt (the teenage years for most people), the body is programmed to grow proportionately, so the growth of tissue and of the skeleton are matched. It stands to reason, then, that excess growth hormone at a young age causes a proportionally large body, a condition known as *gigantism.* In one study of gigantism, a quarter of participants were over six feet six inches, and the tallest participant was eight feet one inch. After this growth phase is complete, however, when the body is not expecting to grow more (at least not upward), excess growth hormone causes a condition called *acromegaly.* The features of this condition are a little different and relate to the fact that the growth plates of long bones are no longer active, meaning the long bones can't get any longer. So acromegaly causes enlargement of the hands and feet. Patients' hands take on a "doughy" feel because of extra soft tissue. Facial features include enlargement of the forehead and jaw. The voice deepens, most likely because of alterations in vocal cord mass and elasticity, and there can be more sweating because of the direct effects of the hormone on the sweat glands. Add this to the features of overactivity of the stress hormone cortisol (known as Cushing syndrome), including weight gain, especially around the midsection, red streaks on the abdomen, high blood pressure, and susceptibility to infection, and you might start to understand why removing a tumor that was producing these hormones would be desirable. And if you're thinking that accessing a pea-size structure right behind the optic nerves deep in the middle of the brain isn't likely to be the most straightforward procedure, then you'd be spot-on. Fortunately, it turns out there is a direct route to the pituitary: through the nose. Or more accurately, through a bone and a cavity, both of which are called *sphenoid.* Transsphenoidal resection of the pituitary, performed using video guidance, is a very effective operation. Ricky's procedure went well, and he recovered quickly.

And so, after three open heart surgeries, one testicular surgery, and one resection of a tiny but powerful tumor deep in his brain removed through his nose, this brave young man was ready for an adult cardiology team.

Ricky was twenty-one when we first met him, and it had been three years since his last open heart surgery. As we all gathered our chairs around his echocardiogram images that day in the clinic, collectively entranced by the bouncing tumor balls inside his heart, we mentally committed to adopting a high threshold for a fourth open heart surgery. Indeed, every open heart surgery makes the next one harder and riskier for the surgeons to accomplish because of the tissue sticking together during the healing process. We decided to follow Ricky every year, measure the size of his tumors, and compare each year's measurements to the last. Over the next few years, they grew pretty slowly but surely. One started to impinge on the mitral valve that lets the blood into the left ventricle, and that was of some concern to us. But his biggest problem, as it turned out, was going to be a tumor we hadn't even noticed yet.

One of the things about ultrasound imaging, of the heart or any other organ, is that some parts of the organ are nicely in focus, but other parts, such as those very near or farther away, are less clear. What we eventually realized, over the course of these years, was that another tumor in the top chambers of his heart was growing much faster than the others. It appeared to have come out of nowhere, though when we went to look back at the old images, we saw that it had probably been there all along, just in the distant imaging field, so harder to make out. Well, it didn't live in the shadows for long. The tumor grew so fast that it started to fill the whole of his right atrium, affecting the functioning of the tricuspid valve that lets blood flow from the top to the bottom of the heart. In fact, to watch that part of his heart was to be reminded of a piston in a combustion engine, since the whole tumor barreled down through the valve and back up with each beat. It quickly became clear something had to be done. Another operation to remove the tumors was possible, but we couldn't just keep doing these surgeries every few years.

I took his case to our weekly clinical meeting, where twenty or so cardiologists and surgeons, along with the nurses, nutritionists, social workers, and trainees, discuss potential cases. The question was simple: Another operation? Or was it time to consider the more serious but potentially curative option—transplant?

The advantage of transplant was clear. The new heart would not have the same genome as his old one, and so it would not grow cardiac tumors. On this basis alone, it was an easy decision. But having someone else's heart implanted inside you comes with a raft of other considerations, not least that many times a day you have to take medications to suppress the immune system. At that moment, Ricky was on minimal medications. In addition, suppressing the immune system would put him at risk for complications such as infections and, well, tumors. Cancerous tumors. He would be trading one known disease for the risk of several new ones.

Ricky seemed wary of transplant but was willing to listen and hear more. After a few months of intensive education and workup that included dozens of blood tests, imaging studies, and other tests, we placed him on the transplant list. Then the waiting began.

One of the things we focused on a lot during the education phase was whether he was ready for transplant. Did he really want it? At times, Ricky seemed nihilistic about his life. Although he had gotten his high school diploma and really loved music, he hadn't yet found his calling, so he spent a lot of time at home in his bedroom. He would play video games, go to bed late, get up even later. He didn't have a lot of friends his own age. When our transplant staff would call, they sensed that, even though he was on the list, maybe he wished he weren't. They asked me to talk to him, and during an emotional call one day, with his mom in the background, we made the decision together to pull him from the active list.

It was a really tough decision for everyone. His family and his medical team were worried about him, worried about his future. We were worried about what would happen if that rapidly growing tumor blocked the blood coming into or out of his heart. Could he drop dead suddenly? Did he understand these risks? We had so many questions, but Ricky was getting really tired of all the questions. He understood that he might die. And he was okay with it.

I remember one clinic visit, I decided to take all the dire warnings, the heart-to-hearts, the medicine, and all the rest, and just put it away. We sat instead and just talked about what he enjoyed in life. Forget all this medical stuff that he'd lived with for so many years. That didn't define him. What did he live for? We talked about music. He played the guitar and liked to write his own music. He explained that, at one point, he had hoped to take music classes at his local community college. His mom and dad were really supportive. But he hadn't gotten around to doing it. Why not? Instead of focusing on living or dying, instead of being asked to make big decisions about surgery or transplant, why not just focus on making music for a while? Get out of the house, go to a class, strum some chords, write some songs.

One night shortly after, I received a call from his mom. She seemed upset and emotional. My mind raced to Ricky, and I pictured the million ways in which three cardiac tumors could have caused catastrophe.

"It's not Ricky," she said.

I was hugely relieved, but also now puzzled. "Ricky's fine?" I asked.

"Yes, totally fine," she said.

"So what's wrong?" I asked.

She explained that her niece, Ricky's cousin, had been found unconscious, rushed to hospital, and had just been pronounced brain-dead as a result of a bleed into her brain. His cousin had no signs of Carney complex, so her death wasn't related to that. In fact, her heart was normal. As it slowly dawned on me the reason for the call, she began to spell it out: his cousin was an organ donor, and the extended family had called her to ask if Ricky wanted her heart.

I was stunned. We are very used to relatives donating a kidney, or bone marrow, or even a piece of their liver to another relative. As long as what remains stays healthy, there are rarely any problems for the donor: one kidney is enough; you have plenty of bone marrow stem cells; and the liver can regenerate. As a result, these living-related donors are common—and they are particularly suitable, because immune matching has a lot to do with how different the genome (and therefore, the immune system) of the donor is from the genome of the recipient. Clearly, you share a lot more genome with close relatives than you do with anyone else.

Obviously, a relative who is still alive is unable to donate his or her heart. In addition, heart donors are so few in comparison to the number of people waiting on the transplant list. For a recipient to even know a donor, never mind be related to a donor, is extraordinarily uncommon. I had never heard of someone donating their heart to a relative.

On top of that, Ricky had already decided he didn't want a heart transplant right now. His mom put him on the phone. A lot was swirling around his head, but in the end, it boiled down to this: to make his cousin's death meaningful, he was willing and ready to go ahead. All the more poignant was that Ricky and his family had just seen this cousin at a wake for her little brother, who had also recently died for an unrelated reason. So much pain for one family, I could hardly bear it. Could some hope be drawn out of this yet?

In the end, for a lot of very complicated reasons, Ricky's cousin's heart did not get transplanted into him or anyone else. But these events served to crystallize in his mind exactly what he wanted: he wanted to live. And with the tumors getting bigger, it was a really important moment for that realization—he was starting to suffer symptoms from obstruction of the blood flow through his heart.

Meanwhile, we were perplexed to not know, even after all this time, what Ricky's disease actually was. It seemed to fit so well with the clinical description of Carney complex and yet that gene, *PRKAR1A,* had been sequenced and looked normal. Could sequencing his whole genome help us find the answer?

It was now 2016, and Stanford Medicine's leadership had decided to build on the research work we had done with the Quake, West, Snyder, and primary care clinic genomes. They had invested in a hospital-based Clinical Genomics Program so that any Stanford patients who needed genome sequencing could have it done quickly and locally. As part of piloting this program, we sent Ricky's DNA for whole genome sequencing at the Illumina lab, then ran the raw data through our variant-calling pipeline in the hospital.

One of our genome curators, Tam Sneddon, took on Ricky's case and started working through the various candidate genes that might explain his disease. Tam was suspicious about the previous finding of a normal sequence for the *PRKAR1A* gene. He wanted to examine that gene closely, and so he pulled up the genome viewer to see the raw data. While most of the gene had good sequencing coverage, there was a strange area near the beginning of the gene, where that coverage seemed to drop. Our computer programs didn't flag that as abnormal, but Tam was suspicious nonetheless.

The intriguing part of it was this: If there was a deletion of part of the gene, that could explain Ricky's classic features of Carney complex (the gene is disrupted) but also the completely normal gene sequence found on earlier clinical testing. (Sanger sequencing only looks to see if a patient's sequence matches what is expected for that part of the gene; it doesn't really pay attention to the number of DNA molecules the signal comes from. If you only have one copy, and the sequence of that copy matches, the result will be a normal test.) Also, Constantine Stratakis had shown, using an older technology called *Southern blotting*, that deletions in *PRKAR1A* could cause the disease. One more modern approach to looking for insertions and deletions in the genome was to use a tool called a *microarray*. Unfortunately, the microarray, which identifies the quantity of DNA in spots scattered sparsely around the genome, didn't cover this region of the *PRKAR1A* gene well at all. We needed to find another way.

A DNA sequencing technology that could perhaps help us came from a Bay Area company called Pacific Biosciences. PacBio (as it is affectionately known) had pioneered a method with distinct advantages. The primary advantage was that it could sequence much longer molecules of DNA. Illumina's technology is an example of short-read sequencing, named because the DNA molecules it sequences are 75–250 letters long. A typical PacBio read at that time, however, was 8,000 letters long. That is like the difference between trying to complete a one-thousand-piece jigsaw puzzle and a ten-piece jigsaw of the same scene. Not only will the ten-piece jigsaw take you less time, but your confidence in the placement of any particular piece at any moment will be much higher. Unfortunately, while exciting and robust, the technology hadn't been able to keep pace with the sharp reductions in cost that Illumina was achieving, so had not been as widely adopted for medical use in humans.

In genome terms, deletions like the one we suspected in Ricky can be very tricky to characterize if your DNA sequence words are only one hundred bases long, and if one of the copies you inherited from your mother or father is normal. So we approached the coinventor of the PacBio technology, Jonas Korlach, who was at that time the chief scientific officer of the company, to see if they could help. Jonas is a biologist with an impressively broad-ranging grasp of engineering, biology, and medicine, not to mention a talented musician who sang for two seasons with the San Francisco Symphony Chorus. Originally from Germany, he invented the PacBio technology with physicist Stephen Turner when they were both graduate students at Cornell. Jonas was enthusiastic and offered to collaborate. The timing was good, because the cost of their technology had recently dropped, and they were starting to think about human applications. Indeed, the company had just sequenced a whole human genome using their latest technology as part of the National Institute of Standards and Technology's Genome in a Bottle consortium. That was a project focused on quality of sequencing, not medical diagnostics. Were we all ready to attempt the first long-read medical genome?

Although the technology was much more mature, and genome sequencing was now widely performed in medicine, at least for rare disease, there were many parallels to the early-days seat-of-the-pants analysis we performed for Steve Quake. For long-read sequencing to find deletions or duplications of whole chunks of DNA, standard bioinformatic tools were not yet available. Fortunately, PacBio had been working with pioneering groups led by Evan Eichler at the University of Seattle and Michael Schatz at Johns Hopkins University to develop new tools in this space. Jason Merker from Stanford and Aaron Wenger from PacBio worked together to conceive and troubleshoot a version 1.0 clinical pipeline.

In Ricky's genome, long-read sequencing identified more than six thousand insertions and deletions (a number that we now understand is pretty standard for the average person). After filtering for previously known common variants that could not cause disease, this was halved and, after prioritizing the list by those variants that would affect the coding portion of any gene, the list narrowed to just thirty-nine deletions and sixteen insertions. We cut the list further, this time only to genes that were known to cause human disease. Finally, we were down to just three insertions and three deletions. We looked at the gene names and immediately one gene jumped out: *PRKAR1A,* the Carney complex gene! The result was undeniable. In one version of Ricky's *PRKAR1A* gene, a large portion—more than two thousand DNA letters—was simply gone. Not there. The benefit of the long reads was clear. The region was covered by some normal reads that clearly came from the intact copy of the gene (and that would have explained the normal Sanger sequencing result) and also by some reads that had two thousand letters missing and must have come from the mutated version. The answer was right there in front of us, with no jigsaw puzzle assembly required. This was an incredible moment.

I remember calling Ricky to let him know that we had finally worked out what caused his disease, more than a decade after his parents raised all that money to pay out of pocket for a single-gene test that didn't find the answer. He sounded really happy. His mom sounded even happier. But despite the joy at this answer, there wasn't yet a curative treatment for Carney complex, and it was becoming clearer and clearer that something was going to need to be done about Ricky's heart.

Ricky was now getting intermittent chest pains and dizziness, which was almost certainly a result of blood being blocked from entering his heart by the tumors. To puzzle out the best solution, we planned to again present his case at our transplant conference, the group who had previously heard his story, approved his listing, then heard back that he just didn't feel ready.

At 7:00 a.m. on a Friday morning, in our cardiovascular building conference room, I laid all the facts of his case out on the table for a room full of surgeons, cardiologists, immunologists, nurses, social workers, nutritionists, and more. I told his story from his first diagnosis and surgery at age seven. I detailed all his surgeries, retold the story about his cousin. I then explained our new molecular diagnosis and how we made it using long-read sequencing. We then discussed what to do.

One thing was clear, though—Ricky still wasn't mentally ready for transplant. He knew something needed to be done, but he wasn't sure he could cope with all of that. If that was the only option, he wasn't ready to take it. I came into the room to make a case: for this twenty-five-year-old kid, getting another five or ten years to live with his own heart, before having to make a truly life-changing decision about transplant, could be a long and valuable time—almost half the time he had already been alive, and double the time he had been an adult in the world. I had a strong feeling that in ten years, Ricky would be a different person. I'd seen him walk a few miles on this journey, consider his own destiny, lose the will to live, and then regain it. He recently had been babysitting for his young niece, and a sparkle would come into his eyes when he talked about her. Suddenly, he could see potential in the world again. In the joy and playful vitality of this toddler, he saw something that he had lost a bit in himself. Perhaps he did want to continue on. Perhaps he could have a family of his own. And while he knew he would ultimately need a transplant, it seemed quite plausible that he would be more ready in a few years. Meanwhile, if we didn't do something right now, he probably wouldn't live long at all. Transplant was off the table. So it was either a risky operation to go in and remove those tumors or go and organize your affairs.

With another regular operation, the group acknowledged, there was the chance that Ricky's heart would fall apart after the tumors were removed and then need to be reconstructed with artificial material. He already had a patch repair of one of the top chambers from his first operation, but patching the bottom chambers of the heart was a very different consideration. This option seemed too risky. Plus, what was that surgery leading to? What was the aim? To push out transplant for another few years? Why not just transplant now? That was safer, smarter. Get the tumors out and the heart that grew the tissue out with them. Even so, a standard transplant would have its own challenges. The surgeon always leaves some tissue from the top of the old heart, to have something to sew the new heart on to. Yet that was the very portion of Ricky's heart that seemed most capable of generating fast-growing tumors. What if, after transplant, those tumors returned? We even discussed a radical procedure known as *autotransplant:* the complete removal of Ricky's native heart followed by dissection and removal of tumors on the "back table" in the operating room, followed by resuturing his own heart back in. There are patients who have had this done as treatment for a very rare malignant cardiac tumor called a *sarcoma.* The discussion went around and around. Even for a room that had borne witness to the pivotal advances in the first development of heart transplantation, this was a hard problem.

The chief of cardiothoracic surgery, the man who took over the hot seat from Norman Shumway, who had pioneered heart transplantation at Stanford, was Joseph Woo. Joe is exactly what you might expect of a cardiothoracic surgeon. He is tall and moves quickly, deftly. He is also smart, confident, thoughtful, and decisive. Like most surgical chiefs, he is always in one of two uniforms: a suit or scrubs. There seems to be no middle ground. He is either working in the operating room (and not uncommonly covered in blood) or dressed in a beautifully pressed suit, probably of Italian design. At the University of Pennsylvania, before his move to Stanford, Joe had gained a reputation as someone who was willing to take on the hardest cases. He worked hard, he worked fast, and he pushed his team as hard as he pushed himself. One of his specialties is repairing valves, the flimsy structures that let blood in and out of each of the four heart chambers. Most surgeons look to replace these when they are damaged; Joe repairs them.

So what did this man who took on cases other surgeons wouldn't consider, this chief of cardiothoracic surgery at one of the most famous departments of cardiothoracic surgery in the world, think of Ricky's case? Joe stared, as we had, at those bobbling tumor balls in the echocardiogram images of Ricky's heart now dancing on a screen just above his head, heard about his three prior open heart surgeries, heard how

Ricky was previously on the transplant list but asked to come off, heard how his cousin had died and offered her heart to him; he took it all in. As he listened to the story, he shook his head in amazement. The eyes of the room were upon him, and a hush came over the crowd.

Joe looked down, stroked his tie, took a breath. Suddenly, he snapped out of deep thought. "Well, then, let's do it. Let's get it done," he said, referring to the risky operation of going in to remove the tumors.

"When shall we tell him to come in?" I asked.

"Monday. Let's get it done."

I met up with Ricky later that day when he came for the preoperative checkup. I could see he was happy that Dr. Woo was willing to consider a fourth open heart surgery.

Joe knew exactly what he was getting into. With a third "redo" operation, even getting access to the heart and blood vessels was going to be slow. Any open heart surgery causes inflammation that leads to adhesions (the tissues of the chest wall literally stick together). Just one prior operation makes a second operation hard. But three prior operations? And then, having opened the chest and gained access to the heart, that's when the challenges would really begin.

The first challenge was where to put the tubes for the heart-lung bypass machine. During open heart surgery, this machine removes blood from the body, oxygenates it, and then returns it to the body. It literally takes on the function of the heart and lungs during the course of the operation. Typically, the venous blood is removed from the top-right-hand chamber of the heart (the right atrium). That was going to be a problem, because that chamber in Ricky's heart was filled with a giant tumor. So the blood would need to be pulled from a vein higher up. Additionally, the decision was made to complete as much of the operation as possible while the heart was still beating. Typically, the heart is stopped after going on the heart-lung bypass machine, but this beating-heart technique helps reduce some of the risks. Nothing was going to be routine in this surgery. Because of the multiple tumors, the team also used a no-touch technique for the heart, meaning they touched the actual heart tissue as little as possible with hands or instruments. This reduces complications afterward, since the heart doesn't like to be touched. Then with Ricky on the bypass machine, it was time to open the right-top chamber, to take a first look at the biggest and baddest tumor. As they inspected the extent of it, another challenge became clear: it wasn't going to be possible to remove the tumor without reconstructing the vena cava, the vein that carries blood from the body up into the heart. The tumor was stuck to the inside of it. The top-right-hand chamber of the heart would also need to be reconstructed, since the tumor was stuck to that as well. So not only was one of the largest veins that drains blood into the heart being removed and reconstructed, the majority of one of the chambers was being replaced as well.

Joe and the team started planning the reconstruction. The part of the vein that would need to be replaced was right next to the heart. They fashioned a reconstructed vein using tissue initially derived from the outer lining of a cow's heart (bovine pericardium). And then they were ready to remove the large tumor. Carefully, they cut away at its stalk and detached it from where it was sticking to the vein. It was enormous. Despite its size, however, it was not stuck to any other part of the heart, so after dissecting the stalk, it came away quickly, filling the whole palm of the assistant surgeon. The reconstructed vein and atrium were now sewn into place to restore the integrity of the vein-heart junction.

Next, it was time to go after the other tumors. The heart needed to be stopped for this critical portion of the operation. They opened the main artery of the body, the aorta, and peered through the aortic valve to see the inside of the heart's main pumping chamber, the left ventricle. They could see the tumors there, straight ahead. To minimize the amount of actual cutting into the heart, it was necessary to remove those tumors through this valve (sort of like trying to grab your jacket off the hook in the hall through the mail slot in your front door). They would essentially be enacting the nightmare scenario we had hoped to avoid all those years—the tumors breaking free and heading out into the circulation. Thankfully, Ricky was on the heart-lung bypass machine, and this would be safe. The tumor that we had seen flopping about near the exit to the heart was the first to come out. That was the easier one, located right in front of them as they stared through the valve opening. Snip and grab. The other one was harder. It was entwined in the mechanics of the inlet valve to that chamber, the mitral valve. Damage to the mitral valve or its cords would create a new problem. This tumor was also not straight ahead. The challenge in removing these sorts of tumors through this tiny hole is the reason some surgeons resort to autotransplantation. It is much easier to remove the heart from the body, open it up, and remove the tumor under direct vision. But that requires a whole lot of handling of the heart and adds massive risk.

After painstaking work using very fine, careful movements, Joe managed to remove that tumor in its entirety, without harming the mitral valve. Remarkably, Ricky's heart was now free of tumors. It was time to let the blood back in and start his heart back up. Time to take him off the heart-lung bypass machine.

As usual, an electric shock was used to bring the heart back to life. Initially, everything looked good. Normal rhythm. Good pumping. Good flow. But then, all of a sudden, Ricky went into ventricular tachycardia. His blood pressure dropped. Suddenly, everyone was in critical mode. What was the problem? Charles Hill, the anesthesiologist—exactly who you want by your side in an emergency—already had an ultrasound probe down his esophagus. He flipped it on, and real-time pictures of Ricky's heart appeared on the screen. The heart, now back in normal rhythm, was pumping well. But there was new, severe leaking of the tricuspid valve on the right side of the heart, the one that the big tumor had been barreling through like a piston with every beat. The tumor must have expanded the valve opening. Now that the tumor was gone, it was leaking badly. Ricky's blood pressure was very low, and the team needed to make a decision. This heart was not pushing enough blood forward to maintain the viability of his organs. Charles was giving drugs to keep him in normal rhythm, but the valve was a mechanical problem his drugs couldn't fix. Charles looked at the screen and looked at Joe. Joe looked at the screen and back at Charles. If this valve needed repair, that needed to be done immediately. There was nothing to do but to put Ricky back on the cardiopulmonary bypass circuit and go back in to fix the valve.

The second time Ricky was taken off the bypass machine, everyone watched Charles's ultrasound pictures, anxiously holding their breaths. After a moment, they started to exhale. There was almost no leaking at all. Ricky's blood pressure was stable. He was looking good. In fact, he recovered so well that the breathing tube came out even before he left the operating room.

I had seen Ricky on the Friday when he came in for pre-op, but I was in North Carolina on the day of his surgery. Joe called from the operating room with the good news. I was ecstatic. The entire team (as well as the patient himself and his family) had accepted the possibility that he might not make it through this surgery. And here he was, already with the breathing tube out and in recovery. The next day, he was off all drips and out of the ICU.

I got back to Stanford two days later, and I immediately went to see Ricky on the third floor in the postsurgical unit. But his hospital bed was empty. I couldn't find him. I started to panic, an anxious feeling rising in my chest and dipping low in my stomach. I found his nurse.

"What happened?" I stammered. "Where's Ricky?" I asked, fearing the worst.

"Oh, he's out walking around the ward," she told me with a big smile.

I almost hugged her. I don't think I've ever been so relieved. Then I spotted his big, smiling face as he ambled over toward me. I'm not sure

I've seen a patient look that good so soon after a heart surgery. Never mind *this* heart surgery. Six days after the procedure, he went home.

We saw Ricky in clinic a week later, and he looked great. He had recovered well; the symptoms of chest pain and palpitations had gone. He'd been up and about every day, and the minimal swelling in his ankles had already disappeared. Although we didn't have a cure for his condition, the molecular diagnosis meant we knew what we were treating. The future could even bring a genetic cure.

We were almost ready to send him home again when I realized we had forgotten to look at his echocardiogram. So I pulled it up on the screen, and the team gathered around. In the same spot where, five years before, we had first been hypnotized by those alien balls bobbing around inside his heart, what we saw instead was nothing: clear black space. The blood was flowing unimpeded into and out of his heart. Four beautiful heart chambers singing together in perfect rhythm. It dawned on us, we were staring at what looked for all the world to be a normal heart. There was not an alien in sight.

# 15

# River of the Land of Pine Trees

*You cannot be a big person with a small heart.*

—T. D. JAKES

*If I were a pony,*
*A spotted pinto pony,*
*A racing, running pony,*
*I would run away from school.*
*And I'd gallop on the mesa,*
*And I'd eat on the mesa,*
*And I'd sleep on the mesa,*
*And I'd never think of school.*

—BEVERLY SINGER AND ARLENE HIRSCHFELDER,
*Rising Voices: Writings of Young Native Americans* (1992)

It was the sound. It was really . . . like an animal sound, like a moaning and I was trying to talk to her . . . but there was no response."

There are no words that can begin to describe what runs through the mind of a mother who has found her teenage daughter blue and lifeless, lying in a pool of her own spit. A few minutes before, Susan Graham's daughter Leilani had been a normal thirteen-year-old—vibrant, fun, sporty, very much like her parents. Susan's husband, Chris, was focused on the rising tide of personal electronics in kids' lives and a generation growing up increasingly sedentary. Both Mom and Dad were active in coaching and refereeing soccer and basketball, invested in lead-by-example encouragement to be active. But there was also more direct encouragement. There was a system in the Graham household: if you want to watch TV, you also run.

On this particular night, Leilani had been working out as usual on the treadmill in the garage. Chris was out at a meeting. Susan was also supposed to be out, but her plans had changed, so she was sitting at home on the sofa that backed up to the wall next to the garage. Susan heard the treadmill whirring, as usual. Then she heard another sound. It did not, she recalled, sound entirely human. But whatever it was, she was clear about one thing: it was not good. She sprinted to the garage and found Leilani collapsed, unconscious, on the treadmill. Guttural moaning emanated from her slowly writhing body. Susan stopped the treadmill, called 911, and waited. Leilani was breathing but not coming around. When the paramedics arrived, she was still unconscious. Could it be drugs? Could it be a seizure? It wasn't immediately obvious what the problem was.

But at the hospital, it soon became clear. An ultrasound of her heart revealed an abnormally thickened heart. Leilani had a life-threatening genetic disease of the heart muscle called *hypertrophic cardiomyopathy*—and that event was not a seizure but a cardiac arrest.

Hypertrophic cardiomyopathy is a disease characterized by overgrowth, stiffening, and overactivity of the heart muscle. It is named for the elements of the disease first recognized: *hyper* ("over"), *trophic* ("growth"), *cardio* ("heart"), *myo* ("muscle"), and *pathy* ("disease")—overgrowth heart muscle disease. It is a not-so-rare disease that affects at least one in five hundred people in the population. In fact, the earliest insights into hypertrophic cardiomyopathy came in the 1700s from anatomic pathologists, whose newly defined craft was the careful dissection of dead bodies. In probably the first description of hypertrophic cardiomyopathy in 1679, the Swiss physician Théophile Bonet described a coachman who had died suddenly in his carriage as having a heart "larger than that of any bullock."

In the eighteenth century, Giovanni Battista Morgagni, an Italian pathologist, was the first to document other key features of hypertrophic cardiomyopathy, including patchy fibrosis, a phenomenon where connective tissue arises in between the heart cells and starts to push them apart. He also described an obstruction to blood flow leaving the heart, created when the middle wall of the heart, the one between the two lower chambers, becomes abnormally thickened. Around this same time, long before anyone had coined the term *genetics,* Giovanni Lancisi, the pope's physician, described a family with this disease in four generations. Lancisi had been asked to investigate an epidemic of sudden deaths in Rome, and his autopsy reports made frequent reference to cardiac hypertrophy.

The condition was to remain firmly in the realm of pathologists for another two hundred years. Appropriately enough, it was another pathologist, Donald Teare in the 1950s, who articulated our modern understanding of hypertrophic cardiomyopathy. Teare published accounts of his prodigious research in papers with understated titles like "Some Reflections on 25,000 Autopsies." (To put this in perspective, if you worked three hundred days a year and did three autopsies every single work day, that would be almost thirty years of work.) His seminal manuscript "Asymmetrical Hypertrophy of the Heart in Young Adults" was published in 1958 and was the clearest description of the condition yet. It included a detailed description of hallmark pathologic changes. Heart muscle bundles, instead of being neatly and symmetrically stacked, appeared in chaotic orientations separated by connective tissue (something we call myocardial *disarray*—and aptly so, given the disruptive effect it can have on both heart rhythm and our patients' lives). For the first time, too, Teare connected these pathologic findings postmortem with prior medical testing of living patients, suggesting ways of detecting the disease in life. He also laid out the clearest case yet for a familial cause: the hearts of two siblings, who both died suddenly, were found to be identically hypertrophied.

The clues to recognizing underlying hypertrophic cardiomyopathy in living patients that Teare first highlighted in his postmortem studies were built upon by Paul Wood, who pioneered the physical examination for hypertrophic cardiomyopathy. Wood was pale, acerbic, and short, with steely blue eyes. He was born in India to a father who was working there as a district commissioner. He trained in medicine in Australia and New Zealand and made his name in London. In his time, he was arguably the most famous cardiologist in the world. Nervous and tightly wound, he was, to some, egotistical and arrogant; to others, brilliant. He reputedly wrote his classic textbook twice. After the first draft was lost, legend tells that he rewrote the whole thing from memory. He was renowned for his skill in patient examination and described in hypertrophic cardiomyopathy patients a jerky pulse, a "double impulse" of the heart on the chest wall, and a "whooshing" sound heard using the stethoscope (caused by the obstruction to blood flow noted above). Wood, notably, died young in a manner befitting his life. After suffering from what he believed to be indigestion for two weeks, he finally faced the fact that his pain was most likely due to a heart problem and persuaded his secretary to carry out an ECG on him. Presented with the tracing, he diagnosed himself with a heart attack. Although he was then admitted to the hospital and given the blood thinner heparin, he did not survive the event. The master diagnostician had diagnosed his own fatal illness too late.

Meanwhile, the first attempt to surgically relieve the outflow tract obstruction of hypertrophic cardiomyopathy had taken place in the UK. In the US, around the same time, in January 1960, at the National

Heart Institute in Bethesda, Maryland, a young physician named Eugene Braunwald, who later himself became one of the most famous cardiologists in the world, referred a ten-year-old boy with aortic valve obstruction to surgeon Andrew "Glenn" Morrow. Shortly thereafter, Morrow summoned Braunwald to the operating room. Could the cardiologist explain, Morrow asked, why the patient's aortic valve looked entirely normal? Morrow decided to measure the pressures in the operating room and found a difference in pressure in different positions inside the heart. He realized that the patient's thickened heart was causing the problem and decided to excise a piece of the wall of the left ventricle. Following the operation, the difference in pressure halved.

Remarkably, Morrow was later himself diagnosed with hypertrophic cardiomyopathy. Aged forty years, suffering shortness of breath on exertion and near-fainting episodes, he asked his friend and colleague Braunwald, who was by then chief of cardiology at the National Institutes of Health and with whom he had spent years studying hypertrophic cardiomyopathy, to listen to his heart. Braunwald heard a murmur and confirmed what Morrow had suspected: he had "the disease." As Braunwald put it to me, "I think that if you saw this story on TV or in a movie, you would call it ridiculous—those things just don't happen in real life. There are these two men who were very close collaborators, and one of them has the condition they're describing, and the other diagnoses it—it's just too cute to believe." Morrow was a brilliant surgeon but, like Paul Wood many years before, a truly terrible patient, and he declined therapies for his disease, including both the medication whose use Braunwald had pioneered (beta blockers—drugs that stop the action of adrenaline), and the operation he himself pioneered that was named after him: the Morrow myectomy. In a tragic turn of events, Morrow died suddenly at only sixty years old, his heart showing at autopsy the classic findings of the disease: asymmetric septal hypertrophy, cardiac cellular disarray, and fibrosis.

In the 1960s, then, most of what we knew about hypertrophic cardiomyopathy came from either blood pressure measurements inside the heart, operating room observations, or postmortems. However, a major new technology was about to impact medicine and revolutionize our understanding of the living, beating heart: ultrasound. The hypertrophy seen at postmortem could now be measured in living patients. Its progression over time could be quantified year by year. The thickening could be distinguished by its pattern and degree from other causes of heart muscle thickening, such as high blood pressure or athletic training.

Driven by the new imaging technology, and the new computing power to perform statistics, two men on either side of the Atlantic contributed more to our understanding of this disease than any two individuals prior: Barry Maron, from the United States, and William "Bill" McKenna from the UK. Over the next few decades, these two published hundreds of papers that described in exquisite detail the electrical and anatomical characteristics of these patients' hearts. This included a phenomenon unique to hypertrophic cardiomyopathy: the abnormal forward motion of the mitral valve during heart contraction. Formally titled *systolic anterior motion,* it is referred to as *SAM* for short (I once gave a talk about it with a different famous Samuel on every slide). Maron and McKenna also tested potential treatment options in populations of patients, including beta blockers and calcium channel blockers, designed to slow the heart and help it relax. They authored some of the first reports of this disease as a cause of sudden death in young people and created methods for how to predict which patients might be at greatest risk. Through their efforts and those of many others, we had advanced dramatically in understanding the *what* of the disease. But we still didn't know *why.*

Neither beta blockers nor calcium channel blockers were going to be enough for Leilani Graham. After arriving at the hospital following her collapse, and having her condition diagnosed by echocardiogram, she was immediately taken to the operating room. Having just suffered a cardiac arrest at age thirteen, she needed a defibrillator, an implanted device that monitors the heart and can deliver a lifesaving electric shock in the event of a serious arrhythmia. The device would be implanted by Anne Dubin, the pediatric cardiologist we met in chapter 12. "She's gonna be okay," Leilani's dad recalls Anne saying. "But you're very fortunate, because 90 percent of the time, her condition is diagnosed on the coroner's table." He reflected that, just a few short weeks before, Leilani had been on top of a volcano on a school field trip to Mexico. "All I could think of was, what if this had happened then?"

From the moment the heart stops delivering oxygenated blood and extracting deoxygenated blood, brain damage begins. The brain is vulnerable because, like the heart, it is an avid consumer of energy. Although it represents only 2 percent of the body's weight, the brain uses about 20 percent of the body's oxygen. One area of the brain called the *hippocampus,* which helps to form new memories, is particularly sensitive. Its sensitivity to oxygen deprivation is the reason even cardiac arrest survivors who are quickly resuscitated often have short-term memory loss. With just one minute of oxygen deprivation, the stunned hippocampus shuts down. After just a few minutes more, brain stunning becomes brain damage, then that damage becomes irreversible. Toxic metabolic byproducts begin to accumulate: reactive oxygen species like superoxide and hydrogen peroxide start to destroy the membranes of brain cells, from the inside. You have half the chance of survival if you are resuscitated after more than three minutes versus less than three minutes. After five minutes, you would be lucky to survive at all. And unless extraordinary circumstances prevail, no one survives more than eight minutes without a heartbeat.

In the hours and days following, there were some confusing conversations. Leilani came out of recovery and, of course, immediately asked what happened. "We'd explain to her, then she'd drift back off," her dad told me. "Then she'd wake up again: 'What happened? Am I okay?' She'd drift off, and then again, 'What happened? Am I okay?' And I'm trying not to panic. I'm trying to explain this to her."

This lack of short-term memory after cardiac arrest can be very concerning to friends and family. Your loved one can talk, listen, understand, and make perfect sense in immediate conversation, yet have no ability to recall any recent events at all. The memory loss is caused by damage to those most sensitive cells in the hippocampus and goes by the name *anterograde amnesia,* which translates approximately as "forward forgetting." You recall established memories, but you can't pick up new ones. Fortunately, the hippocampus recovers, and in Leilani's case, there were no long-term consequences.

Still, it was a new reality to which she awoke. Leilani would now be protected by her device if her heart should stop in the future. However, our goal was to prevent that type of event from ever happening again. For genetic heart diseases like hypertrophic cardiomyopathy, the first line of defense is avoiding undue stress on the heart—and one of the ways we do that is by recommending patients stay away from competitive sports. For many, this has little impact on their lives, but for high schoolers heading toward college scholarships, or for those who enjoy competitive sport as a central part of their lives, this can be really hard. "That was really difficult for her to wrap her mind around," Leilani's dad recalled, "because part of her persona, part of how she viewed herself, her place in the world, was as an athlete—not an elite athlete—but, you know, because she liked the camaraderie of teams. She liked being on a team."

Yet not even these unwelcome restrictions can always prevent another event. Soon after her ICD was implanted, Leilani had a substitute PE teacher, one who hadn't quite understood the gravity of her condition. The teacher suggested that no, really, she should come and do the traditional mile relay with the rest of the class. "I think I got three-quarters of the way around the parking lot," she recalled. "I remember my vision narrowing, feeling short of breath, legs getting super heavy. 'Okay, something's wrong.' And then I just totally face-planted."

That was her second cardiac arrest, terminated by a shock from her defibrillator. The substitute teacher had to watch as two full-size fire trucks and an ambulance screeched, sirens blaring, into the schoolyard, coming to the aid of the teenager with the heart condition whom she had just persuaded, against her protestations, to run merely one lap around the track. Leilani's classmates, too, witnessed it all. "Yeah, I remember talking to my friends who were really traumatized afterward," Leilani recalls. Over time, though, she and her friends adapted and soon got used to her device. She used to flex her muscles to make it stick out as a party trick.

Three years later came cardiac arrest number three. She and her sixteen-year-old friends were exploring the neighborhood near Leilani's home when they came across a hill of dirt. Leilani's California uniform of shorts and flip-flops was not the ideal outerwear for scrambling a dirt mound, but they decided to climb it anyway. Things didn't quite go as planned. "I remember getting to the top of the hill, turning to a friend, and saying, 'I'm going to pass out. Call 911.'" Her friend did. A long, anxious fifteen seconds went by before she woke up. Her friends were, again, freaked out. But Leilani had other things on her mind. "I remember that the guy I was with, I had a huge crush on," she told me. "The way I fell, I was covered in dirt. And so, I remember asking my [other] friend, 'Can you please brush the dirt off my shirt?'"

Leilani's defibrillator was the result of a long and fascinating history of early attempts to restart the heart. It begins with Peter Christian Abildgaard, born in 1740 in Copenhagen, Denmark. His father was an artist, but young Peter was more attracted to science. When he set off to study chemistry and medicine, he had no idea that he would one day make some of the most fundamental observations regarding the electrical activation of the heart. His prodigious talent was obvious, however, and he was selected to be one of only three students sent from Denmark to Lyon in France to study veterinary medicine so that Denmark could acquire expertise in the management of cattle plague. He returned in 1766 to Copenhagen, where he eventually started a veterinary school that focused mainly on horses, in particular, those at the royal stables. This veterinary work was engrossing to the point that he actually stopped practicing human medicine.

Meanwhile, Abildgaard—a true polymath—continued to pursue his interest in physics and invention. He was fascinated by the fact that humans and animals who died of electrocution from lightning bolts appeared to have no outward or inward sign to indicate the cause of death. So he started planning a series of experiments to determine why. Conveniently, the first generators of electricity, Leyden jars, had recently been invented in the Dutch city of Leiden by Pieter van Musschenbroek. These were glass jars containing water and metal foil, capable of discharging a very high-voltage static charge. They were championed by, among others, Benjamin Franklin. So naturally, Abildgaard built his own and, in a breathtaking experiment that would pass no modern ethics committees, tried electrocuting horses by delivering electric shocks to their heads.

To his surprise, Abildgaard found that no amount of electricity could successfully slay a horse. He did manage to stun one, but to his disappointment, it got up shortly thereafter. Concluding that "with no amount of effort could I kill this Herculean animal," as he wrote in a later paper, he "reluctantly" moved on to a nearby hen, wrestling it to the ground. To this poor bird's head, he applied the same shock from his jar that had, minutes before, been enough to stun a horse. His hen collapsed, lifeless. Morbidly excited at this scientific "advance," Abildgaard moved to deliver a second shock across its temples, fervently hoping that this would resurrect the animal. It did not. Indeed, as he later reported, the animal remained dead "even after repeated shocks to the head." Frustrated and defeated, his hopes of scientific immortality fading, he decided to try one last electric shock, this time through the chest. *Zap!* The animal promptly opened its eyes, stood up, and "walked quietly on its feet."

Hard as it might be to believe, from this cartoonlike scene would emerge a pivotal insight into our understanding of the heart. Like any good scientist, of course, Abildgaard needed to repeat the experiment. So he delivered another shock to the head of the resuscitated hen. The bird fell down dead, again. After a fourth shock, this time through the chest, it was back. Resurrected twice in a row! Even deities hadn't achieved that feat. Delighted, Abildgaard began the cycle again. In his "classic" hen reanimation paper, published in 1775, he wrote, "After the experiment was repeated rather often, the hen was completely stunned, walked with some difficulty, and did not eat for a day and night." But, remarkably, after that, the hen appeared "very well" and even "laid an egg."

By the late 1770s, the power of electrical resuscitation had moved beyond chickens, and there were now reports of electric shocks delivered to humans who had "apparently" died—with miraculous results. The first of these appeared in a report of the Royal Humane Society, and described the recovery of a person who had "apparently drowned." Another report, from the Royal Society in London, described a three-year-old girl, Sophie Greenhill, who fell from some height and also appeared to have died. She was, in fact, pronounced dead at the scene. Nonetheless, an apothecary was called, and he decided to try electricity. Although a full twenty minutes had elapsed after the "apparent death" before he was ready to deliver the first shock, and although he wasted quite a lot of time shocking various body parts other than the heart with no effect, when he eventually applied shocks to the chest, all of a sudden, Sophie recovered a weak pulse. To everyone's astonishment, a few minutes later, she started to breathe. After ten minutes, she sat up and vomited. Much like Abildgaard's hen, she was disoriented and shaky for a few weeks, but she made a complete recovery.

Word of miraculous resuscitations such as these captured the imagination of the public and physicians alike. One British physician, James Curry, started to gather these examples into a handbook. Published in 1790, *Observations on Apparent Death from Drowning, Suffocation Etc* became a guidebook to the use of electricity for resuscitation. It contained nuggets of therapeutic gold such as this:

> *When the several measures recommended above have been steadily pursued for an hour or more without any appearance of returning life, Electricity should be tried; experience having shewn it to be one of the most powerful stimuli yet known, and capable of exciting contraction in the heart and other muscles of the body, after every other stimulus has ceased to produce the least effect. Moderate shocks are found to answer best, and these should, at intervals, be passed through the chest in different directions, in order, if possible, to arouse the heart to act.*

History would judge James Curry two hundred years ahead of his time. Those statements about the electrical excitement of the heart and the need to pass the resuscitating current in many directions through the chest were preternatural. Curry's advice was now broadly actionable, too, thanks to a near portable device invented by his contemporary Charles Kite. In his 1788 "Essay on the Recovery of the Apparently Dead" (they don't make titles like they used to), Kite described how the device harnessed static electricity from Leyden jars. Reanimation by electricity soon became known as *defibrillation,* a term first coined by Edmé Vulpian in French literature, and derived from the Latin word *fibrilla,* meaning a small fiber (in this case, a heart muscle fiber). Although the medical profession itself was slow to realize its importance, society at large took an increasing interest. The most famous example in popular literature comes from Mary Shelley's 1818 novel, *Frankenstein.* The protagonist, a young scientist named Victor Frankenstein (contrary to popular misconception, it was the scientist, not the monster, who was named Frankenstein) becomes fascinated by the secret of life and sets about gathering body parts from various places with the idea of stitching them together to create a new life. Searching for the ignition trigger to spark this new being to life, he comes upon electricity: "I succeeded in discovering the cause of generation and life; nay, more, I became myself capable of bestowing animation up on lifeless matter."

The 1800s saw the first truly portable defibrillators, but society still wasn't ready to trust this "sorcery" entirely. Citizens wary of being experimented upon by "vigilante galvanists" started sewing labels into their clothing with requests to be left "unelectrified" if they were to be found unconscious. Even in our own times, when patients are advised to place a detailed advance directive in their medical chart regarding their wishes if their heart should stop, some choose to leave nothing to chance. A case report in *The New England Journal of Medicine* described a seventy-year-old man who arrived at an emergency room in Miami unconscious with a Do Not Resuscitate tattoo emblazoned across his chest at the exact spot you might place the pads for medical defibrillation. After an ethics consult, they did indeed accede to his wishes.

After a fallow period in the early 1900s, 1947 brought another landmark: the first medically documented internal defibrillation during an open heart surgery. When a fourteen-year-old boy's heart fibrillated during surgery, the surgeon was first forced to perform *cardiac massage*—a procedure whose soothing name is wildly at odds with its accompanying frenzy. You take the heart in your hand and literally squeeze the blood out. Then repeat. Until you have a better plan. While manually pushing blood out of the boy's heart provided temporary relief, it didn't solve the problem. The heart continued to fibrillate. Eventually, as a last-ditch effort, the surgeon decided to try electrical defibrillation. The heart stopped, paused, then promptly started beating normally. The operation concluded, and the boy recovered completely.

External defibrillation—the application of electrical energy across the chest from the outside—took a little longer to reach prime time. It was significantly advanced by a man who would later become the dean of engineering at Johns Hopkins University in Baltimore, an electrical engineer from New York named William B. Kouwenhoven. And if you are right now wondering how to pronounce his last name in your head, don't worry: he was known as "Wild Bill" to his friends. He kept his equipment, the product of a decade of research, in a locked lab on the eleventh floor of the Johns Hopkins Hospital. One day, his medical resident, Gottlieb Friesinger, finding a patient in the emergency department on the ground floor to be in cardiac arrest, ran up to the eleventh floor, persuaded a security guard to open the lab, carried the equipment the eleven floors back down to the emergency department, and performed the first successful, medically documented, external defibrillation on a patient in cardiac arrest.

Wild Bill was also part of the group at Johns Hopkins who developed *cardiopulmonary resuscitation*, or CPR, the combination of chest compression and artificial respiration (blowing air into the lungs of the collapsed victim). Although techniques for artificial respiration had been reported for decades, and maybe centuries, it was in the 1960s that the medical community first came to formalize the combination of mouth-to-mouth resuscitation and chest compressions as an integrated method.

Even if your own heart is incapable of pushing out blood by itself, with effective CPR, you can stay alive for hours. Yet effective CPR (push hard and fast) and—more important—immediate CPR, rarely happens outside of hospitals. If you are out and about in the world and your heart starts to fibrillate, what you need, frankly, is an electric shock. This is what those automatic external defibrillators (AEDs) visible in airports, gyms, and hospitals are designed to provide and why these are some of the safest places in the world to have a cardiac arrest. You might be surprised at another safe place. Can you think of somewhere else with a strong interest in keeping its patrons alive and also keeping a watchful eye on those patrons? It turns out, that one of the places you are most likely to strike it lucky if you have cardiac arrest is a casino. A famous study published in *The New England Journal of Medicine* described the impressive lifesaving benefits of a program to deploy defibrillators in casinos after training the staff to use them. Speed matters.

Those portable devices in stadiums and casinos owe much to Paul Maurice Zoll, the son of Lithuanian and Belarusian immigrants, who, by the 1970s, had devoted much of his career to making devices to save patients from arrhythmic sudden death. His path was defined in his final year as a medical student at Harvard University, when he became profoundly affected by a patient he lost. Although many cardiac arrests are caused by fibrillating hearts, where uncoordinated electrical activity leads the normally pumping muscle to simply vibrate ineffectively, some cardiac arrests are caused by a more fundamental failure of the electrical system: the heart flatlines with no electrical activity at all. In those situations, delivering a shock is minimally effective. What you need instead is *pacing,* where the delivery of rhythmic shocks of lower intensity causes normal heart beating to resume. In essence, you replace the biological pacemaking function of the heart, normally assumed by a group of cells at the top of the heart that pass their electrical activation down the electrical circuits of the heart to drive a heart contraction. At this point in the 1970s implanted pacemakers had been around for some time and could perform this task for the patients who had them. The question Zoll was focused on, however, was: What if the patient in front of you suffers a cardiac arrest and needs pacing but doesn't yet have an implanted pacemaker? He realized that in addition to delivering one high-energy shock from a defibrillator to bring the heart out of fibrillation, you could also use the same apparatus to deliver lower-energy pacing shocks across the chest. Although understandably rather painful for the patient, who (unlike a patient in ventricular fibrillation rhythm) is often conscious, this technique continues to be used to this day and can manage life-and-death emergencies until a pacing wire can be advanced into the heart via a vein in the neck. Zoll's name remains familiar to cardiologists today through the ZOLL Medical Corporation and their almost ubiquitous portable devices for delivering defibrillation and pacing.

Placing defibrillators close enough to be moved quickly to someone in need still likely leaves a brain precious minutes without blood. What if there was a way to deliver help faster?

This was the professional obsession of Mordechai Frydman, a Polish Jew whose mother died of heart failure when he was young. He left his home in Warsaw at age fifteen to escape the Nazis and, likely as a direct result of that decision, was the only member of his family to survive the war. Renamed Mieczysław Mirowski in Nazi-occupied Poland, he showed the dedication, stubbornness, and unbounded optimism not untypical of those who succeed despite all the odds. "Even then, I felt that I could overcome all difficulties," he said of his early exit from his Warsaw home. "This was, and is, an irrational idea. In retrospect, it seems completely crazy that at the age of 15, I should leave my father, my town and my country." He went on to train as a doctor in France and Israel and landed as director of a coronary care unit attached to the Johns Hopkins School of Medicine in the United States. In this role, Michel, as he became known, was used to dealing with cardiac arrest. But it was one cardiac arrest in particular that sealed his life's obsession. "In 1966, my old boss, Professor Harry Heller, started having bouts of ventricular tachycardia," Mirowski said, as John Kastor reported in *The*

*American Journal of Cardiology* in 1989. "He was repeatedly hospitalized and treated with quinidine and procainamide. My wife asked me why I was so concerned. 'Because he will die from it,' I told her." And he did, two weeks later while having dinner with his family.

Mirowski, despite having none of the electrical engineering background you might think you'd need to attempt the feat of miniaturizing a defibrillator so it could be implanted *inside* a human, nevertheless dedicated himself to exactly that. There were minimal funds, many rejections, and countless naysayers. One in particular, Bernard Lown, at the time one of the world's foremost experts on sudden cardiac death, wrote, "In fact, the implanted defibrillator system represents an imperfect solution in search of a plausible and practical application." But the path of history is not marked by the tread of those who give up easily. And you don't make it out of Nazi-occupied Poland at age fifteen without an unusually potent fighting spirit. Mirowski had demonstrated proof of principle in multiple dogs, but he continued to push up against the conservative cardiology community and an industry reluctant to disrupt its consistent profits from a thriving pacemaker business. With single-minded determination, Mirowski, pushy and persistent, and by now balding and bespectacled, forged ahead. In 1972, he persuaded the CEO of a small firm called Medrad to dedicate engineers and resources to miniaturizing his device to be fully implanted in a dog. That CEO, Stephen Heilman, devoted three years of his research-and-development budget at Medrad to achieve that aim. When they finally managed it, they made a video of the device in action. The video is not publicly available, but Paul Wang, the chief of adult electrical cardiology at Stanford, has seen it, and he described it to me. A dog is seen running around wagging its tail and eating a tasty snack. A shock is delivered manually from the device with exactly the right timing to stop the heart. The dog promptly stops running and collapses, crumpling into a heap on the floor. Then, silent and unseen, the implanted defibrillator device detects the fatal rhythm, diagnoses it as fibrillation, summons the battery, builds a charge, and delivers a shock to the heart. Pause. The dog opens its eyes, gets to its feet, and starts running around again. All without human intervention. The sight of a dead animal being brought back to life without a human touch—waking up, standing up, running around, and wagging its tail—captured the imagination of the medical community in a way that a written report never could. The direction of the wind changed almost overnight, and in 1980, after several more years of refinement, Mirowski's device, weighing 280 grams (just over half a pound), was implanted in the abdomen of a human for the first time. Electrical wires ran from it to patches that were placed on the surface of the heart. The device, an implantable defibrillator, or ICD, received FDA approval in 1985.

Videos of actual humans being defibrillated back to life are equally captivating. One famous example, available on YouTube, shows Anthony Van Loo, a professional soccer player with an inherited heart disease, suffer a cardiac arrest on-camera during a game. You see him initially collapse to the ground like a rag doll. At the moment he hits the grass he has, in a very real sense, died. His heart is fibrillating. No blood is moving around his body. And though you can't hear it, there is a clock ticking down the minutes to irreversible brain damage. After a few seconds, almost before anyone on the field notices he is down, and before anyone gets anywhere close to considering CPR, his whole body convulses for a split second. His implantable defibrillator just fired. It detected the fatal rhythm within milliseconds, charged within seconds, and delivered a lifesaving shock. A moment later, he sits up. Just like Abildgaard's hens. Just like Mirowski's dogs. Before the team doctor, who rushes onto the field to stretcher him off, has even arrived, you can see him negotiating to stay in the game. This is not uncommon for a patient whose ICD restores normal cardiac rhythm during unconsciousness before the brain really knows what has happened. They are out for those seconds and don't feel the shock. They wake up a little disorientated—often wondering why they are on the ground, since *they* don't remember falling. *They* do things like request to stay on the soccer field after they just *died* and were brought back to life in front of thousands of people. Despite his pleas, the paramedics politely declined Van Loo's request and took him to the hospital.

Leilani's defibrillator had brought her back twice, and each time, she had woken up disorientated. This was a feeling she recognized. But despite these life-threatening events, she continued to thrive. Although she had, at a younger age, defined herself in relation to team sports, she adapted to her new constraints, taking an increasing interest in the arts and particularly in music and acting. As she moved toward high school, she chose choir over more strenuous activities, and in the spring of her senior year, she was accepted to both the acting and musical theater programs at New York University (NYU). She chose musical theater.

I first met Leilani around this time, when she was seventeen. We were the adult clinic and the time for transition had come. Time to leave the playfully decorated walls of the pediatric clinic and the doctors with the Donald Duck ties to enter the monotonically beige world of adult medicine. "There are fewer fish on the ceilings," she would later wryly observe. To help with these transitions, Heidi, my nurse colleague, and I would go over to the children's hospital to meet the patients on their turf. We never really know what to expect in such meetings, but even we were surprised at our first meeting with Leilani to encounter such an assertive young woman. "Leilani ran the meeting," Heidi recalls. We didn't realize at the time what an impact this remarkable young woman was going to have on us.

In addition to her three cardiac arrests, one of the things we discussed at that first meeting was genetic testing. At the time, genetic testing was pretty new for hypertrophic cardiomyopathy, and we were one of the few clinics in the country to offer it. We suggested to Leilani a test that included about a dozen specific genes that were known to cause the disease. She had recently taken an advanced placement biology class and had brought some notes and some very good questions. In the end, after a description of what the test could offer and what it might and might not show, she decided she wanted to be tested.

Hypertrophic cardiomyopathy is a disease that can result when just one of the two copies of a gene is affected. For this reason, it is called a *dominant* disease (the mutated copy of the gene "dominates" the normal one). In most cases, that disease-causing gene is inherited directly from the patient's mother or father—which means that parent is also at risk to develop the disease. There is a fifty-fifty chance for each of our patient's brothers or sisters, too, that they inherited the disease-causing gene. That is why family screening is so important. Often, the patient's mother, father, or siblings are already affected without even knowing it. So in addition to performing genetic testing for Leilani, our highest priority was to screen her mom, dad, and little sister, Kai, for the disease using ECG and echocardiography and—because the disease can develop with age—to plan for future screening.

The good news was that her sister's heart looked good on ultrasound. That was a relief, but it didn't mean she was in the clear forever. She could still have inherited a family predisposition to disease but might not show signs of hypertrophic cardiomyopathy until much later. That kind of variability is not uncommon in families. We also screened her mom and dad, and neither seemed to show any overt signs of hypertrophic cardiomyopathy, despite the fact that, most likely, one of them carried a genetic variant that underlay Leilani's condition. If we looked really, really closely, perhaps there was a subtle hint in her dad's ultrasound. I remember thinking his heart looked like it was trying just a little too hard (one feature of the condition is that the heart contracts more vigorously than it should), but this was not definitive. So perhaps the genetic testing could help us here. If it could identify the disease-causing variant in Leilani, we would then have a yes-or-no test for other family members, capable of telling us who was actually at risk. We sent the test off and waited.

This was 2009, and genetic testing remained an expensive and slow operation. To sequence one gene required amplifying each protein-coding section (or exon, typically a few hundred DNA letters) in a specific reaction then taking that DNA and sequencing it using the Sanger method described in chapter 1. This was followed by comparing the gene sequences for each segment of each gene to the reference sequence for that gene. This was a huge amount of work, very little of it automated. Variations found in Leilani's DNA, but not in the reference, would then be checked against a panel of one hundred or so anonymous blood donor controls. If a new variant was not found in any of those seemingly healthy individuals, it was reported along with a best guess as to whether that variant might actually cause the disease.

The first clue in the hunt for the genetic cause of hypertrophic cardiomyopathy came from an unexpected place. Next to the sweeping lush green Parc Laurence in southeast Québec, a few miles from the American border and about twenty miles south of the nearest large town, Sherbrooke, is the town of Coaticook. The name Coaticook is of Abenaki origin, meaning "river of the land of pine trees." In French, the area is known as Perle des Cantons de l'Est because of its natural beauty, but Coaticook became known for something far less beguiling: the Coaticook "curse" that took many of its citizens early with strokes and sudden death. In the fall of 1957, two brothers, aged thirty-nine and forty-one, originally from this town, were admitted to the Royal Victoria Hospital in Montreal with strokes. Both were found to have enlarged hearts. Both also came under the care of Peter Paré, a six-foot-four lung specialist known for his gentle, humorous teaching style. Fascinated by the coincidence of strokes and enlarged hearts in the two brothers, Paré became convinced there must be a genetic explanation. So much so that he dedicated the next few years of his life to understanding the cause of disease in this family. The brothers, he learned, were not alone. In total, he studied seventy-seven family members, asking many to come to the hospital for clinical examinations, some of them several times. In July 1961, he published his seminal paper, "Hereditary Cardiovascular Dysplasia—a Form of Familial Cardiomyopathy," in which he described twenty living affected family members as well as another ten who had previously died, and even traced the origin of the disease back to one particular immigrant who came to Québec in 1650. In fact, in the four years it took him to complete the work, five family members died suddenly. If ever there were a family waiting for the right genetic technology to narrow down the cause of a condition, it was this one.

Genome-scale sequencing, of course, was a long way off. The Human Genome Project wouldn't get started for decades. But the linkage/mapping technology was invented in 1980, and it could help. With this technology, and a family like the one from Coaticook, the pieces were in place. The seminal work describing the genetic underpinnings of hypertrophic cardiomyopathy was led by a husband-and-wife team, Christine and Jonathan Seidman of Harvard Medical School. Together with Bill McKenna in London, they approached the family first characterized by Dr. Paré and invited them to weekend "reunions," where they could learn about this research study and consent to join. More than one hundred family members signed up for what was to become a tour de force of genetic discovery. The linkage-mapping method was labor intensive but extremely powerful. So powerful, in fact, that the Seidman laboratory estimated odds of two billion to one that the gene responsible was in the identified narrow region. Given strong odds that this was the right corner of the genome in which to look, the team started fine mapping by sequencing the critical genes in this area until they found one gene, the cardiac myosin heavy chain (*MYH7*), part of the molecular motor of the heart, that looked to be responsible. The genetic variant was found in every single family member with the disease. The Coaticook curse turned out to be one misspelled letter in the genome: where there was supposed to be a G, there was an A.

Reported in 1990, the discovery was seminal. It made hypertrophic cardiomyopathy one of the first genetic diseases to have its cause clearly identified. Given the development of lifesaving defibrillators, this was also a finding that, along with family genetic screening, could save lives. And not unlike striking oil in a new region of the world, finding the first-known genetic cause of this disease helped direct researchers where to look for others. The following years led to the discovery of even more disease-causing variants in the *MYH7* gene (including one affecting the family originally described by Donald Teare in the 1950s), as well as variants in additional genes that can also cause the condition. Even better, little by little, hundreds of years after this disease was first described, the true cause of hypertrophic cardiomyopathy began to be unveiled.

When the results of Leilani's own genetic testing came back a couple of months later, they were not at all what we were expecting. In Leilani's DNA, we found *two* genetic variants in a key hypertrophic cardiomyopathy gene, *MYBPC3.* Myosin-binding protein C (the gene I saw on the screen in Steve Quake's office that first day we met) is a key part of the molecular motor of the heart. It binds to the protein product of *MYH7,* the myosin heavy chain, the first gene shown to cause HCM, discovered by the Seidman laboratory. In Leilani's case, both variants seemed independently capable of causing the disease. This was important news in so many ways. First, it helped confirm the actual cause of her condition. Heart muscle thickening and arrhythmias can have many causes, and while the shape of her heart was strongly suggestive of hypertrophic cardiomyopathy, until we knew the genetic cause, we could never be entirely sure this was the correct diagnosis. The test also helped to clarify why her presentation was so severe: she had two variants. Certainly, two disease-causing variants would go a long way to explaining her dramatic presentation at a young age. What remained puzzling was that neither of her parents seemed to be affected. If these variants were so bad, why weren't her parents showing signs of the disease? Perhaps the variants were totally "new"—and instead of being passed on by a parent, they arose in Leilani for the first time? In the language of genetics, they would be called *de novo.* We thought that unlikely, because we knew that de novo variants causing hypertrophic cardiomyopathy were rare (and, as we learned in chapter 6, each of us has only about fifty new variants in our entire genome). It was more likely each of her parents carried a variant but, unlike Leilani, who had received a double dosage, her parents didn't manifest signs of the disease. To find out, we immediately tested Chris and Susan for the two variants found in Leilani.

As it turned out, Leilani had indeed received one variant from her mom and one from her dad. And, true to the subtle abnormalities on his echocardiogram, the gene variant that came from her dad appeared to be a larger contributor. This clarified that both her mom and dad were at risk for hypertrophic cardiomyopathy themselves, and both needed to be seen regularly in clinic going forward. That left her younger sister. Although Kai's recent echocardiogram and ECG had been normal, this did not mean she was in the clear. She could still have inherited one or both of the disease-causing genetic variants and be at risk of later developing the disease. Without the predictive power of a genetic test, she faced years of continued screening with a cardiologist, including an ECG and echocardiogram annually during her teenage years. After that, even if her heart still looked normal, she would need to transition to the adult hospital and be screened every five years. Or she could take the genetic test and know for sure whether all this was necessary. The only real question was: Did she want to know?

Having witnessed what happened to her sister, Kai told us she wanted the testing. She understood the sobering odds: 75 percent that she was at risk for the disease. There were two chances in four that she inherited just one disease-causing variant from one parent; one chance in four that she had both variants; and one chance in four that she was free of risk, having inherited neither variant at all.

So we sent off the test and waited. The result came back a few weeks later. On the four-sided die, Kai had thrown a zero! Neither variant had been passed on to her, meaning that she was at no more risk of this disease than a random person walking down the street. She also could not pass the family disease on to her children, because she didn't carry those disease-causing variants in her body.

We often worry about the complexity of emotion that shared genetic risk can foment in a family, when the play of chance means that risk is not distributed evenly. Leilani had been severely affected by her hypertrophic cardiomyopathy, while we now knew that her sister had escaped any risk at all. Sometimes, as well as a sense of unfairness in the affected, there can also be survivor's guilt in the unaffected. Parents are challenged with the idea that they have passed on disease-causing genes that have caused their child to suffer. It's especially complicated when the disease is one the parents don't appear to suffer from themselves. Of course, in families with genetic disease and in families without, we inevitably pass on to our children both our "good genes" and our "bad genes." We can never forget that it is the child's response, the environment they grow up in, and their resilience in the face of life's challenges that will define them just as much as, if not more than, their genes.

Leilani's resilience would soon face its biggest test, as became apparent in the fall of the next year when she started school at NYU. Moving your daughter out of the family home to college is clearly one of the hardest and proudest moments of any parent's life. There is a particular poignancy, though, in dropping your daughter off across the entire country in New York City after she has survived three sudden cardiac deaths. "My dad cried," Leilani noted. "I had only ever seen him cry three times in my life."

When classes began, the challenges were immediate. NYU had a strict policy on lateness, and Leilani had trouble convincing the administration that the long distances between her classes were going to be a challenge since her heart condition meant she couldn't speed walk or run. A physical disability from a failing heart, she noted, was treated differently from a brain or spine condition, where a wheelchair or mobility aid might make it more apparent to outside observers that transitions would take time. She managed to make it work during the sunny, crisp days of fall.

"Then the snow came. It was awful."

One afternoon, not far from Greenwich Village, Leilani was coming from a voice lesson. She had been feeling the cold and was a bit out of breath, so she would usually have taken the elevator. Instead, she followed her voice coach up the stairs where they finished their conversation and parted. Moments later, she collapsed, limp and lifeless, on the busy sidewalk.

This was cardiac arrest number four. And once again, the defibrillator saved her life. "I remember waking up feeling like I had woken out of a dream, because I was staring up, and there was sky and trees. The sun was setting." People were crowded around, some kneeling at her side. Others just kept on walking. After a few more minutes, she was once again quite alert. Not unlike soccer player Anthony Van Loo, who tried to negotiate staying on the field after his cardiac arrest, Leilani was all about minimizing fuss in the ambulance. "I remember there was a paramedic around my age," she told me. "I was like, 'Just let me out. I'm just going to take a taxi.'"

"No, that's not how this works," he had replied with a smile.

She texted her parents from the ambulance. "Hey, don't worry. But I'm in an ambulance."

Shortly after, she flew home. Her mom and dad suggested—tentatively—that maybe she should consider not going back to New York to finish three more weeks of classes and take her finals, to which she replied, "Hell no. I just killed myself this semester. I'm definitely going back." The irony of her choice of metaphor was not lost on anyone.

# 16

# Songs in the Key of Life

*You very soon in your mind realize that it's not just*
*An ordinary pain in your heart.*

—STEVIE WONDER, "ORDINARY PAIN," FROM *Songs in the Key of Life*

*So, though this was absolute shit, for now: I choose joy.*

—LEILANI GRAHAM, *A Calculated Risk* (BLOG)

After her cardiac arrest on the streets of New York, Leilani returned to school at NYU and completed the semester as she had promised, before eventually transferring to the theater program at the University of Southern California. No. More. Snow. The new climate and relaxed lifestyle were exhilarating, and she was so much closer to home in case of emergency.

As time went on, however, she began to notice new symptoms: a lower blood pressure on her home monitor, occasional light-headedness, more trouble exerting herself on the stairs. Most patients with hypertrophic cardiomyopathy live a pretty normal life, limiting themselves from competitive sports, and perhaps taking some medication to control symptoms. Some suffer more and require protection with a defibrillator or relief through surgical removal of thickened muscle. A smaller number still suffer much more—with dangerous heart rhythms or a heart that stiffens or weakens and eventually fails, requiring a heart transplant. Leilani was at risk of this happening, but it was not clear when that might be. It was something that lived in the future, that indistinct, out-of-focus place you reasonably don't spend too much time dwelling on when you are getting on with living your life. After one trip home and a visit with our team when we first brought up transplant as a real possibility, she remembers heading back to Southern California, thinking that it still "didn't feel real." She was just out of shape—a different exercise regimen, a new diet, a personal trainer, perhaps? Any of those could get her back to where she was a year or two ago, couldn't it? It couldn't be that her heart was getting worse. It was close to college graduation, so it couldn't be time to talk about transplant, could it?

Medically, though, it was pretty clear the direction we were heading. Tests showed that Leilani's heart had stiffened and weakened enough that it wasn't pushing sufficient blood around her body. The time had come to make the decision about whether to pursue a transplant. Before agreeing, Leilani wanted to hear from more doctors. We encouraged her to chat with David Rosenthal, her pediatric cardiologist. David is the other half of one of the most dynamic duos in pediatric cardiology, as he is married to Anne Dubin. Anne had taken care of Leilani during her initial admission and placed her defibrillator, and David was the expert in hypertrophic cardiomyopathy and had been seeing her regularly until we took over (we cardiologists divide ourselves into electricians, plumbers, and mechanics—David is, like me, a mechanical cardiologist focused on the heart muscle itself). After graduation, Leilani had started working at Google, which offered employees access to a "second opinion" service. In total, four cardiologists reviewed her case.

In October of 2015, we presented her story to the forty or so members of the Friday morning transplant meeting. Leilani was accepted for listing with no hesitation, and the waiting began.

Transplant can be an incredible gift. To have the heart of another person beating inside you is a miracle of surgery, extending and validating the life, and the death, of the person who donated it. But waiting for that transplant is excruciating. Your name is on a list. You don't know the length of the list. You don't know anything about who else is on it. And you never know exactly where you are in line. You wake up every day and try to get on with life, knowing that, at any moment, you might have to drop whatever you are doing and drive to the hospital for a surgeon to saw your breastbone in two and slice out your heart. Plus, the longer you wait, the sicker you become. It therefore comes with a strange emotion: the uncomfortable, ambivalent reality of having to "hope" someone else might die a premature death so you can live. "I wanna say like my mind was like 75 percent transplant and 25 percent work," Leilani told me. "I mean it was just, you know, constantly in the background. And I felt like I wanted more information, but I didn't wanna know too much. I wasn't totally ready for it, but I was also so tired of that feeling of waiting. I felt like I had put my whole life on hold. I did think that part was really hard."

Four months later, Leilani was up at her family cabin in Volcano, California, watching her dad through the window as he burned some brush at the bottom of the hill. That's when her phone rang. "I remember the guy on the other end had this superlong introduction, 'Hi, I'm calling from Stanford Department of Cardiothoracic Surgery. . . .' I was like, 'Just say it!' Oh, my god, it took forever. I was standing, and I was pacing, but I was like, I should just sit down because I felt light-headed and just sort of 'on' and, like, flooded with adrenaline." This was *the call.*

"I just kind of collapsed on the stairs and just cried vehemently for like thirty seconds and then stopped. And Mom is yelling down to Dad, and he's trying to put out the bonfire he'd been kindling. I think we packed up and left in about five minutes."

We normally ask our patients not to stray more than a few hours from the hospital, because as soon as a heart is allocated, a dance begins with multiple surgical teams coordinating around the donor patient to retrieve the lifesaving organs. We don't take the transplant recipient to the operating room until we have eyes and hands on the donor heart, but we like to have the recipients come to the hospital as soon as possible so they are ready in place.

After a frantic drive, Leilani showed up at Stanford University Medical Center:

"Hi, I guess I'm checking in."

"And what are you here for?"

"A heart transplant."

Although friends, family, and the health care team all provide support in difficult times, groups of patients suffering the same disease can provide each other uniquely individualized support through shared experiences. One of the places Leilani found support over the years was the Hypertrophic Cardiomyopathy Association. The HCMA serves over twelve thousand families and is led by HCM patient, tireless advocate, and fearless crusader Lisa Salberg. Lisa, with flame-colored hair and a feistiness of character befitting her Irish heritage, lights up any room. She was diagnosed at age twelve, and not the first in her family, either: her grandfather and uncle both died of sudden cardiac death in their forties; she lost her great-aunt to a stroke, possibly related to HCM; and, closest of all, her sister, Lori, died when she was thirty-six from HCM. For Lisa, the disease is not a hobby or a job; it is her life. She founded the HCMA in 1996 to support, educate, and advocate for patients and families with HCM. Even through hospitalizations for stroke, placement of a defibrillator, heart failure, and her own heart transplant, she was to be found communicating with patients just like Leilani. Now, with a new heart beating in her own chest, powering her indefatigable spirit, Lisa carries around a plasticized version of her old heart: a reminder of the life she lived and a disease she personifies but that does not define her. With a flourish, she will often pull the model of her old heart out of a metal lunch box, to the amazed gasps of those around her. To hold her heart in your hand is to understand the resilience of those who survive with a pump that is overpowered yet underperforming. You stare at it and can't escape the metaphor of a new heart and a new beginning.

A few hours later, renowned cardiothoracic surgeon Phil Oyer carefully removed Leilani's old, stiff, failing heart and started to sew in a new, soft, supple one. Her old heart was intercepted, as with all our heart transplants, by my research team, who took it to our lab to extract tissue samples for analysis of its RNA and proteins before forwarding it on to our pathology department for a clinical report.

Before the procedure, Leilani, an actress and singer, had given strict instructions to anyone she encountered to be careful with her vocal cords when placing the breathing tube required during anesthesia (and this significantly understates the force with which she conveyed this sentiment). A few hours later, the breathing tube was gently removed. For patients whose operations go smoothly, and who wake up quickly, especially the younger patients and those who walk in for their transplants (as opposed to being wheeled in from a critical care unit), there is a moment before the strong immune suppressant drugs kick in when things can appear almost too normal. Yes, you are in the intensive care unit, and yes, there is a new heart beating in your chest, but you are awake, you made it through, and suddenly, there is a lot more blood coursing through your veins than there was yesterday.

When I went to see her, Leilani was already sitting up awake, reading in bed. For the first time since her initial cardiac arrest when she was thirteen years old, there was no defibrillator under her skin. She remembers this period, too, then adds, "And that's the last memory I have of things going okay."

The first I heard of things not going okay was later that day. Leilani's new heart had skipped a few beats. Then out of the blue, it paused for a brief moment. No big deal. Unusual in a new heart, but definitely nothing to worry about in isolation. That heart, after all, had just been taken out of one body and implanted in another. But then it happened again, and we started to get more concerned. We grabbed an ultrasound machine to take a look at the new heart beating inside her chest. And then, the new heart paused again. This time the pause was long, too long. A code blue was called. We all ran to the bedside. Since Leilani lost consciousness, we put the breathing tube back in, preparing for uncertainty ahead. I looked at the monitor. The heart rhythm looked fine now. Definitely a bit slow, but normal. Her blood pressure and oxygen levels were now in a good range. We looked at her labs, at the acid level in her blood. Everything looked fine. Except that the heart, her new heart, had briefly just stopped and restarted. We stood around her bed anxiously watching the monitors: her surgeon, her anesthesiologist, several nurses, and me. We were puzzled, worried, anxious. What was happening?

We ran through the possibilities. This heart had looked perfectly normal before it was taken out of the donor. It looked and felt great in the hand, according to her surgeon. There were no issues with the surgery and, when the time came to electrically restart the transplanted organ, it started right up with no problem. True, the right side of the heart was not pumping that well, but that was very common after transplant. The left side of the heart was pumping perfectly. It was not being rejected by the immune system. There was no evidence of blockage in the coronary arteries. The valves were not leaking. This seemed to be an issue isolated to the electrical system. Some local infection, perhaps, affecting the conduction system? Some detail of the donor's medical history not detected by standard screening procedures?

We were preparing a watch-and-wait strategy, since the electrical system seemed to have recovered. Then it happened again. Right in front of us. The heart stopped. The regular blips on the screen disappeared, leaving a straight, flat line. We couldn't wait any longer. We had to save

Leilani. We decided to use a specialized kind of heart-lung bypass machine called ECMO—*extracorporeal membrane oxygenation*—literally, oxygenating blood outside of the body. But oxygenation wasn't why we needed it. Leilani's lungs were fine. We needed it because its pump could take over the circulation and because it could be inserted *right now*. Although we can insert the blood tubes for ECMO at the bedside, Leilani was thirty feet from the operating room door, so we wheeled her directly there as her surgeon scrubbed.

The whole thing happened very quickly. When Leilani's mom and dad had stepped out thirty minutes prior to go get coffee, Leilani was doing fine. When they came back, it was to the lights and noise of the code blue. *Oh shit, that's her,* Chris remembers thinking. *That's not good.*

The medical team, in emergencies like this, is fueled by adrenaline and purpose. There is no time for emotion, only for thinking as fast as possible and for action. Gather information, think through scenarios, talk out loud, argue the odds with each other, come up with some plans, decide on a single plan, implement the plan. We crowd out emotion with the fierce urgency of analysis and action. The family, on the other hand, have the adrenaline, an overwhelming anxiety, and instead of action, only waiting. Painful, nauseating, seemingly interminable waiting. Leilani's parents were turned away from her room and ushered into a small, stark family waiting room around the corner. At one point, a nurse entered, checking to make sure they were okay. "Are you here to prep us for the fact we've lost our daughter?" Chris asked in a panic. The nurse didn't have much information, only that Leilani was still alive. "I thought we'd lost her that time," her dad later told me. "I thought for sure that she was gone."

It was a long wait. It was evening now, and not knowing how much they had been told, I went to find them. Hospitals can be depressing places at night. Around intensive care units, with mortality rates of 10–20 percent, there is a lot of grief; rotations of tired relatives; and rivers of tears. I found Chris and Susan at the quieter entrance, just outside our echocardiography lab. The windows there look out over the Stanford campus, now enveloped with a dark blanket of cold foreboding. We hugged, and I explained what was going on and why we had made the decisions we had. I tried to be positive. But there's no easy way to say, "There might be a problem with your daughter's new heart. We're putting her back on a heart-lung machine until we know more." I tried to suggest they should get some sleep. But none of us were going anywhere.

Heart transplants are typically the final resort after a long road of therapies aimed at reducing the burden of disease. However, none of these myriad therapies were developed from an actual understanding of the underlying cause of the disease. In fact, despite major breakthroughs in our understanding of the genetics of cardiac hypertrophy in the early 1990s, it remained unclear how those gene variants actually *caused* the disease. Where did the hypertrophy, the connective tissue, and the overly powerful contractions actually come from? And why were there dangerous rhythms?

To answer that, we turn to Stanford scientist James "Jim" Spudich. Jim had been considering how to cap off a glorious career studying molecular motors that included a Lasker Award (generally considered a warm-up to a Nobel Prize). He had contemplated, and indeed, we had asked him many times to consider studying the molecular motor of the human heart. Unfortunately, the exquisite molecular tools he had invented required large amounts of human protein that were challenging to obtain experimentally. Fortunately, this was an area that his close friend and colleague had been working on since the 1980s.

Leslie Leinwand, a brilliant scientist and beloved mentor to generations of trainees, has been studying hypertrophic cardiomyopathy for many years. She made many seminal discoveries, including on the importance of diet, hormones, exercise, and metabolic factors in the expression of cardiac hypertrophy in males versus females. In recent years, she has turned her attention to studies of cardiac growth in well-fed pythons (pythons increase the size of their heart by 50 percent in the two days after one of their rare, but often gargantuan, meals). She had also been working on methods to generate enough human myosin, the cardiac muscle motor, to allow it to be studied in the lab—just the sorts of quantities that Jim Spudich could use. But this proved to be more problematic than anticipated. The first obvious option would be to purify the myosin proteins from diseased human hearts with mutant proteins. That was impractical since those tissues are so scarce. The next approach would be to make large quantities of those proteins in bacteria or other easy-to-grow cells and organisms. Several investigators took that approach for years and were unable to obtain functional proteins. In the end, the missing insight was that this type of myosin needed to be made in muscle cells. Leslie cracked it. Suddenly, she had oodles of functional human myosin. This was the key advance that would allow her and Jim to progress our understanding of one of the most important myosins of all: the one in the center of the human heart.

Jim is tall, with a gentle demeanor that belies his imposing scientific presence. A beard frames a huge smile. He is as warm and generous a human being as you could meet, with a wonder for science undiminished after a long and prolific career. In one technique he pioneered, myosin molecules are stuck to the surface of a glass slide, and actin filaments with fluorescent "light bulbs" attached are added. These actin filaments are like train tracks being moved around by an upside-down locomotive. As soon as fuel is provided for the myosin motors, they start to grab the actin filaments and pull on them, moving them around in various directions across the slide. Under the microscope, these appear like little glowing "worms" zigzagging around the screen. It's truly wondrous. In another technique, called a *laser trap*, an actin filament is pulled tight between two laser beams and a single myosin molecule is brought up toward it, where it then pulls on the filament. Force from a single motor molecule can then be measured by extrapolating from the amount of energy needed by the laser to hold the actin filament in place.

Using tools like these, Jim started to work on the question of what it was about the mutated genes in hypertrophic cardiomyopathy that really caused the large-scale problems of heart muscle thickening, stiffening, and hypercontraction. At first, it seemed obvious. We know hypertrophic cardiomyopathy is associated with greater force of contraction, so we'd expect that if you make a myosin molecule identical to one in a patient with hypertrophic cardiomyopathy and measure its force of contraction, you would find a stronger molecular motor, right? Not so fast. As Jim made and tested more and more of the myosin mutations found in our patients, that was not the pattern he was seeing. In some cases, especially the mutations found in children, the force was indeed greater, but in some it was unchanged, and in others, the force measured from the mutant protein was actually lower. What was going on?

Sometimes in science, breakthroughs come from unexpected places. Jim had been studying myosin for many years. His family knew just how much his obsession with molecular motors had contributed to science. But sometimes, they just wanted him to switch off. It was on just such a night, in fact, in December of 2014 that his wife, Anna, suggested he stop worrying about why the myosin force measurements weren't adding up and instead read a book. She even gave him a science fiction novel to help: *Haunted Mesa* by Louis L'Amour, set in the American Southwest. This wasn't a random choice. Jim is a longtime pilot and has enjoyed many hours flying in his Piper Comanche around the western United States, especially around the Southwest's mesas. Flying he loves; science fiction novels, not so much. Twenty pages in, he was sound asleep.

He woke at 5:00 a.m., out of a dream about mesas and, of course, myosin. He dreamed myosin itself had a mesa—in other words, a flat area with steep sides and that this area was particularly important to myosin's function in a way that we had not yet understood. He jumped out of bed and immediately pulled up the myosin structure on his computer. And there it was: a flat-topped mesa, its structure almost identical in dozens of different species. There and then, he overlaid onto myosin's three-dimensional structure some of the well-known mutations that cause hypertrophic cardiomyopathy. Many of the mutations fell directly on the mesa. Even more remarkably, the kinds of genetic changes that caused problems in that area suggested that this was an area of the protein that another protein might stick to. By now, it was 6:00 a.m. His mind was racing. What other protein might attach here? One in particular came to mind: myosin-binding protein C. This was the gene I first recognized in Steve Quake's genome file in his office that first day we met and the same gene harboring the two variants that caused hypertrophic cardiomyopathy in Leilani. Jim also thought of one other possibility: it could be another part of the myosin molecule, bending back on itself.

Jim called me a little bit later, still excited. He had a new idea, he said. Could I meet for coffee? I looked at my schedule. My morning was full, but I didn't hesitate. This was not an invitation I was turning down. I canceled my meetings and enjoyed one of those incredible moments in science when you get to be close to something—and someone—truly special. I sat back in the cafeteria of our medical school's Li Ka Shing building, bright light shining through the floor-to-ceiling windows as he laid out his new theory. The myosin's function was being modulated by other proteins, like myosin-binding protein C, binding at the myosin mesa. Somehow this binding was changing the "pulling" action of myosin. After he finished, I picked my jaw up off the floor and asked a simple question: How could we help? Could my lab, for starters, prove that this myosin mesa was indeed a place where patients' disease-causing mutations were more likely to cluster? Jim eagerly agreed, and we started planning.

To work out if genetic variants were overrepresented in the mesa region, we needed two things: genetic data on a large population of patients with hypertrophic cardiomyopathy, and genetic data on a large population of people without hypertrophic cardiomyopathy. We had recently joined with other large cardiomyopathy centers in the United States and Europe to build a patient registry run by my colleague and friend Carolyn Ho at Harvard University. We also called on another friend, Daniel MacArthur, who had built and released to the public a searchable database containing genetic information from tens of thousands of individuals not known to have hypertrophic cardiomyopathy. Finally, we called on one of our lab graduates, Rick Dewey, who was by now with the Regeneron Genetics Center, who helped us access variants from another large database of people without HCM from Geisinger Health in Pennsylvania. Together, we had data on genetic variation in the myosin gene from more than one hundred thousand individuals. One of my graduate students, Julian Homburger, was eager for the challenge of testing Jim's "mesa dream theory." We asked the question: Are there any areas of the myosin molecule where patients have many more variants than non-patients? One new advance was to model the areas of the myosin protein in three dimensions rather than just in the genetic code so that areas that were actually next to each other in the final molecule were considered together—kind of like looking for the weak spot in a thread if you suspect your shirt button might come loose. If you look at the thread wound through the button hole, one frayed area that might lead you to lose the button is obvious in a way that it wouldn't be if you were to unwind that same thread out on a table (the equivalent of only looking at the genetic code alone).

After running the models, Julian found two different areas where disease-causing variants clearly tended to cluster. One was previously guessed at, but never proven. It was the "hinge" region of the myosin, a region called the *converter*, because it converted energy to movement. The other? The other was an area at the top of the molecule. More specifically, it was a flat area with steep sides. The computer model had found the myosin mesa! This was a formal confirmation of Jim's idea and the first time I had completed a scientific experiment to test an idea that came to a collaborator in a dream.

Meanwhile, Jim's team was puzzling over how mutations at the mesa surface led to the increased force of cardiac muscle contraction that is seen in hypertrophic cardiomyopathy patients. His thought was that, at any one time, a certain number of myosin motors are actively pulling on actin and so are contributing to the overall force, while some are kept away from actin by being folded back onto their own tails, something modulated perhaps by the myosin binding protein. Sometimes, we explain this to patients by asking them to imagine a rowboat. Each myosin molecule has oars that stick out that can be used to power that boat across the water. The more oars in the water, the more force generated to move the boat. This has, in fact, been my favorite analogy since one of my physiology premed lecturers, Neil Spurway, leaped onto a table in the lecture hall at the University of Glasgow in front of 250 students to demonstrate myosin "rowing through the actin sea." In Jim's model, some oars are in use, but others are held back "in the boat" and are not available for pulling. The myosin binding protein appears to hold some oars out of the water. What then if mutations at the mesa meant the myosin binding protein couldn't bind as tightly, so was no longer as good at its job? What if, in other words, the disease-causing genetic variants increased the number of "oars" that were actually in the water? Remarkably, Jim's recent work has shown exactly that. As a result of these findings, and similar observations from myosin scientist Raúl Padrón at the University of Massachusetts Medical School and the Seidman laboratory at Harvard, it now seems increasingly clear that the increased heart contraction in hypertrophic cardiomyopathy is caused by a shift in the balance of myosin heads that are available for pulling versus those that are not. While much work is left to be done to connect this mechanism with all the downstream effects of hypertrophic cardiomyopathy, we now have a glimpse of how the genetic changes first described almost thirty years ago might actually translate to cause the disease in patients.

Leilani made it to the other side. The new heart seemed to stabilize and started to recover its pumping function. The valve on the right side was leaking less. The rhythm was normal. It remained very unclear what had happened, but every day was better than the last. We tried to make sure one of our team visited most days. Some days it was me, some days our lead nurse, Heidi, some days our talented cardiology fellow, Victoria "Vicki" Parikh, would go. When Leilani faced another complication requiring a visit to the operating room on one day, the room was full—Heidi, Vicki, Susan, the ICU physician, and the anesthesia team—and they did a power chant, willing her forward. On another day, it was Leilani and her family. She was, slowly but surely, detached from the machines and drains. Soon, she was out of intensive care and on the regular ward. The exit door was, ever so slowly, coming into focus.

Unbelievable as it may seem, this wasn't the only medical challenge the family was facing at this time. A few weeks prior, Leilani's dad, Chris, had quietly let us know he'd been feeling palpitations, and so we sent him for an echocardiogram. You might remember that he carried the more severe genetic variant of the two Leilani inherited but that his heart had looked relatively normal on ultrasound when we first looked at it years before. Also, he had never had any symptoms. High levels of stress and the accompanying adrenaline, however, are not good for hearts prone to arrhythmia. When the call came for Leilani that a heart was available, Chris had to put the brush bonfire out, pack up the cabin, and make the three-hour drive to the hospital. Then there was the terrifying wait, the overwhelming sense of relief as she came out from the operating room looking good, and the crushing fear when her heart stopped and he thought he was going to lose her. As his daughter lay there in the intensive care unit, results from his own rhythm monitor came back, and it became clear we were not going to avoid having a different conversation. We had to talk about an implantable defibrillator for him.

Chris's heart monitor showed a few beats of ventricular tachycardia. Although structurally his heart appeared relatively normal, with its walls not noticeably thickened, we knew better than to ignore these few rapid, chaotic beats. We knew about his genes and his family history. Specifically, a daughter with multiple sudden death episodes, and in that context, this rhythm was not only meaningful, it was actionable. True, his risk was likely lower than Leilani's—since she had inherited two variants that, together, caused a particularly early and severe form of the disease. But with clear evidence of ventricular tachycardia in Chris, the conversation couldn't wait. We talked about it outside the ICU while Leilani was lying inside, and later my colleague Matt Wheeler saw Chris in clinic. As Chris put it, "When Matt said to me, 'You are at risk of sudden death,' it's kind of like, 'Okay, got it, risk of sudden death.' And it was funny, because at the time, he said, 'Don't you have any questions?'" Given everything he had seen Leilani go through, Chris's answer was simple.

"No, not really."

Over the following months, we remained puzzled about what could have caused Leilani's sudden downturn in the immediate aftermath of her transplant. But since her new heart was working well and there was no sign of transplant rejection, we put it down to some unusual quirk of the stress of the transplant. She had other things on her mind. The high doses of steroids that we give to suppress the immune system after transplant had caused damage to the bones of her knees, making running impossible. This was a particularly hard blow since, for the first time in years, her heart was ready and willing to power her up hills and down into valleys, wherever she would want to go. She poured her emotion into her writing and started to increase her patient advocacy work, sharing her story to help others.

But now and again, she would experience a familiar sensation: a palpitation, or a transient feeling of dizziness. At first, we thought it might be her blood pressure or the effects of some of her medications, but we realized that we had to investigate. My friend and colleague Marco Perez, the electrical cardiologist who sees our adults with inherited arrhythmia (and whose back I am used to seeing disappearing away from me uphill on one of our regular mountain bike rides), placed an implantable recorder the size of a paper clip under her skin to monitor every heartbeat. We really didn't know what to expect. Nevertheless, the result surprised us. It showed evidence of heart block and even a brief run of ventricular tachycardia. How was that possible? Could this have been related to the failure of the heart shortly after transplant? While the cause was unclear, it was quite apparent what needed to be done. A pacemaker or defibrillator needed to be implanted. So Marco placed a pacemaker, close to the spot that the old defibrillator had been. Leilani was devastated.

We were still trying to reconcile, on the one hand, a heart that looked and contracted perfectly normally and that had suffered no rejection episodes, with the electrical faults we were now witnessing. The donor had not, to our knowledge, had any prior heart problems, but we don't always have a full medical history on donors. Was it possible? Could it even be conceivable that this new heart was also affected by a genetic disease of the electrical system? I spoke on the phone to Leilani as she sat waiting for her pacemaker implant. She asked me the same question and asked whether we might genetically test her new heart. While it seemed incredibly unlikely by probability alone, we could think of no better explanation. There was only one thing to do. We decided to sequence the genome of her new heart.

However, sequencing a heart is not quite as easy as it sounds. There was no guidebook for carrying out genome sequencing on a transplanted organ. We weren't even sure it had ever been done. Normally, we would sequence a person's DNA from a blood sample. Yet a sample of Leilani's blood wouldn't be helpful, since it would contain her own genome rather than that of the new organ. What we needed was a heart sample. Fortunately, heart transplant patients get biopsies of their new organ several times in the first year after implantation, to make sure it is not being attacked by their immune system (the process called *rejection*). We aimed to isolate DNA from this tissue and sequence that. Our Clinical Genomics Program took on the challenge, and a few weeks later, we had a list of the genomic variants present in Leilani's new heart.

When Megan Grove, our genetic counselor whom we met in chapter 7, scanned that list of variants, she found something startling: there was a mutation in myosin-binding protein C, the same gene that had caused Leilani's hypertrophic cardiomyopathy. She did a double take. This would be remarkable. Leilani's new, transplanted heart also had hypertrophic cardiomyopathy? Then, however, Megan realized what must have happened. This wasn't just any variant in myosin-binding protein C, it was one of *the very same variants* found in Leilani's old heart. Suddenly, it made sense. When we had isolated a piece of her new heart, there were clearly also some cells in there (most likely blood cells) that belonged to Leilani and contained her own genome. We had unknowingly sequenced two genomes mixed together, the new genome and the old one. Clearly, we needed to adjust our computer program so it would pick out just the genome of her new heart. To do that, we had to sequence her original genome on its own, from a new blood sample, then "subtract" the sequences from the file so we could look just at the DNA from the new heart.

A few weeks later, I met with in my office with Megan and Liz Spiteri, the lab director of our clinical genome program (and talented metal sculpture hobbyist) to discuss the results. We were joined by the lead genetic counselor of our Center for Inherited Cardiovascular Disease, Colleen Caleshu. Colleen is a force of nature. Just before she started with us in 2011, she began training for a full Ironman (about ten to fifteen hours of running, biking, and swimming, if you're counting). She brought that same discipline and intensity to her academic and clinical work, along with intellect, compassion, and a natural leadership style characterized by a gentle, self-deprecating humor. She proceeded to assemble an incredibly talented group of cardiovascular genetic counselors around her. At last count, there were ten. Colleen had not been with us when we first met Leilani and discussed her *MYBPC3* variants, but Colleen had gotten to know Leilani over the years since, discussing issues like family planning and helping with testing the extended family. Suddenly, now with a new heart and new genome, there was even more to talk about.

In the cardiac gene list, one variant in particular seemed to stand out. It was in a gene called *connexin 40* (*GJA5*). Connexin proteins help to form connections between cells called *gap junctions.* In electrical tissues, like the brain or heart, these gap junctions are particularly important for electrical conduction. In one family with a variant in this gene, there was progressive block of the heart's electrical system. Also, mice that were engineered without this particular gene had problems with electrical conduction and bad rhythms. In other words, a mutation in this gene would be a strong candidate to cause exactly the sort of electrical problems that Leilani's new heart was experiencing.

We couldn't say definitively that this was the cause, but clearly we needed to discuss the finding with Leilani. Colleen called her and arranged for the three of us to meet in my office. We wondered how she would react. Since the time she got her pacemaker (and maybe even before), she had come to strongly believe that there was something wrong with her transplanted heart. She had researched online to find as much information as possible about the donor. She had studied rhythm abnormalities. She herself had asked if we would consider genetically testing her heart.

After Colleen described what we had found, we both looked at Leilani to gauge her reaction. "So it's not definitive, but this could be the cause of my new heart's problems?" she asked.

"We think so," Colleen replied.

Leilani broke into a smile. "I knew it."

"You seem happy?" I asked, a little surprised.

"Yes, I was right," Leilani said.

This was confirmation—tentative confirmation—of what she had suspected from the beginning: after all that she had been through, receiving two genetic variants from Mom and Dad that both predisposed her to the same disease, after multiple cardiac arrests, followed by heart failure, and even a heart transplant, she had received another heart that appeared to suffer from a genetic disease.

Leilani has begun a new journey as a patient advocate, blogging and speaking about her experience. Her natural ability at performance, combined with her astonishing life story, makes every word compelling. Her accounts of living a life so entwined with the medical system are stirring, heartfelt, gut-wrenching, and achingly beautiful. In music, too, her voice is serenely beautiful. I've had the privilege of accompanying Leilani on piano as she performed vocals at two of our patient events. She moved a room to tears once with a haunting rendition of "Over the Rainbow," a performance all the more poignant for the parallels with Eva Cassidy, the young singer-songwriter whose arrangement we performed, who also faced daunting medical odds and who died of malignant melanoma at age thirty-three.

It's hard to put into words the feeling you get when a patient triumphs over adversity—thinking about what fate doled out, what burden such young shoulders have had to bear. Leilani Graham has lived through more at twenty-six than most of us will in a lifetime. She walks a line between, as she describes it, “a sort of gratitude” and “for lack of a better word, anger.” And although she is still writing the pages of her own life story, her courage, her resilience, and her fierce advocacy on behalf of patients like herself is an inspiration for us all.

Part IV

# PRECISELY ACCURATE MEDICINE

# 17

# Superhumans

*In a world of ordinary mortals, you are a Wonder Woman.*
—QUEEN HIPPOLYTA, *Wonder Woman* (1975)

*A true hero isn't measured by the size of his strength,*
*but by the size of his heart.*
—ZEUS, *Hercules*

There was no question Eero Mäntyranta was one of the finest Olympians ever to grace the podium. He grew up poor in rural Lapland, an area known for its snow and reindeer, in the cold northwest of Finland. He first stood on skis at three years old. For him, cross-country skiing was a close to year-round mode of transport, not a leisure-time winter sport. The only direct route to his elementary school was across a lake, and the journey, much of it completed in the dark in winter months, could take an hour. In pushing his cardiovascular system to propel his lean, muscular body on rickety wooden skis several miles across a frozen lake can be found the origins of a story so remarkable that the Olympic gold medals he was destined to win don't begin to tell the tale.

Mäntyranta started winning cross-country ski events at an early age, first locally, then at regional events, and then nationally. His Olympic debut was at Squaw Valley, California, in 1960, where at twenty-two years old, he won gold in the 4×10 km relay. By 1964, he was in peak condition and dominated the 1964 games at Innsbruck, Austria, so convincingly that they nicknamed him "Mr. Seefeld," after the Alpine town where the cross-country skiing races were held. It wasn't hard to see why: He won the 15 km race by more than forty seconds, an unprecedented margin. In all, over a twelve-year career, he won seven Olympic medals, three of them gold. His achievements were recognized by the Finnish Ministry of Education, and a museum dedicated to his athletic prowess was erected near his home in the town of Pello.

And yet Mäntyranta's career was dogged by rumors of blood doping. A form of performance enhancement used by athletes to cheat, blood doping involves removing a liter or two of your own blood months prior to an event. The red blood cells that carry oxygen are removed, and the plasma that is left over is reinfused into your body. Your body senses the loss and starts making new blood cells. Meanwhile, the removed red blood cells are stored until just before competition, then reinfused, providing an artificial boost. The boost comes from the fact that extra red blood cells carry extra oxygen to the tissues. The same idea actually underlies the legal technique of altitude training (and possibly some of the remarkable performances by female athletes shortly after giving birth when they still have the higher number of red blood cells they had during pregnancy). At altitude, the lower oxygen pressure in the air stimulates the bone marrow to produce more red blood cells in response to a hormone secreted mostly by the kidney called *erythropoietin*, usually shortened to *EPO*. In the modern era, some athletes—one of the best known being American cyclist Lance Armstrong—inject themselves with an artificial form of EPO to stimulate red blood cell production. The challenge for the sports authorities with blood doping is that since it is your own blood that you are injecting back, there is no "artificial" agent like an anabolic steroid or EPO to detect. So to detect doping, the test is a simple as this: How much of your blood is made up of your red blood cells?

This number, known as the *hematocrit,* can be derived quite easily: just spin your blood in a centrifuge to separate out the three components—

red cells, white cells, and plasma. Normally, 35–45 percent of the volume of your blood is taken up with the red blood cells. Altitude training could push that up by a few percentage points, perhaps to the high 40s. In an athlete, we would generally suspect blood doping if the number were above 50. Above 55 would be very unusual.

Mäntyranta's numbers? Sixty to seventy percent.

It is hardly surprising, then, that rumors of blood doping abounded. Even his insistence that he'd had such high levels measured as far back as his teenage years did not seem to convince many people. However, when several other members of Mäntyranta's family were found to also have high percentages of red blood cells in their circulation, his family began to attract the attention of Finland's hematologists and geneticists. One geneticist in particular, Albert de la Chapelle, had made a name for himself in identifying the genes underlying many family-based conditions using the linkage mapping technique we described in earlier chapters. He started collecting DNA from as many of the family members as he could. On one particularly fruitful occasion, de la Chapelle visited Mäntyranta's home, where he had gathered forty family members. As the geneticist later told David Epstein, author of the book *The Sports Gene,* he was, at one point in the evening, speaking to three older women seated on the couch and discussing their health, when he realized that the only one of the three who was describing any health problems at all was the one without the mutation. A mutation that appeared to be associated with long life (many relatives were elderly and still in good health) and exceptional physical performance? What was this?

De la Chapelle eventually evaluated ninety-seven members of the Mäntyranta family across five generations, of whom twenty-nine were affected by the condition. Many family members didn't even know they were affected. Next, he and his scientific team set out to investigate the cause. Before getting to genetics, they examined Mäntyranta's EPO levels, expecting them to be high. They were surprised to find the opposite: Mäntyranta's numbers were at the low end of normal (normal values are between 8 and 43, and Mäntyranta's was 8.6). They next looked at a sample of his extracted bone marrow and found it to be hyperactive at producing red blood cells when stimulated with EPO. In fact, they found his bone marrow was active even *without* the usual need to add external EPO to the dish. What was going on? It turned out Mäntyranta's bone marrow was exquisitely sensitive to EPO. Even the tiny amount extracted from his body with the bone marrow sample was enough for it to continue to make red cells in the dish.

Now they turned to the genetics. They measured markers spaced across the genome in each family member and found one specific marker that was shared by all the affected individuals. Moreover, the marker was right next to the gene for the EPO receptor, the molecule that senses the presence of EPO and helps carry out the directive to make more red blood cells. In the affected members of the family, the mutated receptor acted like a car accelerator pedal that was stuck in the down position, constantly stimulating the marrow to make red blood cells. In fact, among the affected Mäntyranta men, the proportion of blood volume occupied by red blood cells was 60 percent.

Twenty years after Mäntyranta's retirement from competitive skiing, this genetic finding was at last sufficient to acquit him. In his fifties at that time, Mäntyranta was nevertheless grateful that the shadow of suspicion was finally lifted. And although he himself would always point to his training and what he believed to be his psychological edge (both of which clearly were sine qua non), there remains no question that his superhuman mutation was a key contributor to his athletic success. As further evidence of that genetic advantage, consider the other members of his clan. There was Pertti Teurajärvi, Mäntyranta's nephew, a double Olympic medalist in cross-country skiing. Also, his niece was a national champion.

One study, in fact, showed that pushing the percentage of red blood cells to just 50 percent by giving EPO to athletes boosted maximal performance by 10 percent—a huge margin among elite athletes who are often separated in their performance by just milliseconds. More dramatically, the time to exhaustion at a given level of exertion was increased by up to 50 percent. Just the sort of advantage that would help an elite cross-country skier separate himself from the best in the world by more than forty seconds.

Mäntyranta's remarkable story led us to wonder: What other superhumans might be living undetected among us?

At Stanford, we have a superhuman study inspired by Eero Mäntyranta: we call it *Exercise at the Limit—Inherited Traits of Endurance*, or *ELITE* for short. Our aim is to study the genomes of the fittest athletes on the planet to try to discover some of what makes them so exceptional. In this case, we define fitness not by how many gold medals someone has won, since that involves a complex interplay between physical and mental fitness, but rather by a measure known as *VO2max*, the maximum amount of oxygen that an athlete can extract from the air. To determine an athlete's VO2max, we exercise them to maximum intensity, usually on a treadmill, but optimally using whatever training modality they are most used to (sometimes a bike or rowing machine). At their maximal exertion, we measure the oxygen content of the air that the athlete breathes out. Since we know the oxygen content of air at sea level is a stable 21 percent, we can calculate how much oxygen they extract. This number is highly predictive of performance in endurance events. It does not explain the whole story—biomechanical efficiency and an ability to deal with the buildup of metabolic byproducts of exercise are also very important. But VO2max is the most universally accepted, easily measurable indicator of fitness. People, of course, vary widely in their numbers. Normal adults can extract about 25–35 ml of oxygen per kilogram of body weight per minute (although it's a gas, oxygen is measured as a volume in liters). Serious recreational athletes can often get up into the 50s. Most people measured in the 60s or 70s have competed at a national or international level in an endurance event. Those with numbers higher than that are in a tiny superhuman category that contains fewer than one in ten thousand people. The cutoffs for inclusion in our study are 75 ml/kg/min for men and 63 ml/kg/min for women. Many athletes, especially those in ball sports like soccer, American football, or basketball, never reach these kinds of levels. In fact, it is possible to be a multi-gold-medal-winning endurance athlete and not qualify for our study. Ryan Lochte, the eleven-time Olympic-medal-winning American swimmer, has a publicly reported VO2max of 70 ml/kg/min—extremely high, but not high enough for our study. Lance Armstrong would have made it, likely even without the added boost from blood doping or EPO: he recorded 85 ml/kg/min. One of the highest numbers ever recorded was from one of the most decorated winter athletes of all time. Bjørn Dæhlie, a Norwegian cross-country skier who has won twelve Olympic medals and seventeen world championship medals, clocked in at 96 ml/kg/min. According to his physiologist Erlend Hem, this was an off-season measurement, meaning midseason he might have exceeded 100 ml/kg/min. The highest recorded measurement is from Oskar Svendsen, a Norwegian cyclist and cross-country skier who tipped the scales at 97.5 ml/kg/min.

In my lab at Stanford, the ELITE study is led by Swedish physiologist and international man of action Mikael Mattsson. Tall, blond, and with a muscular frame he carries in an unassuming manner, Mikael occasionally sends us photos back from the Belize rain forest or some jungle in Paraguay, where he is participating in a grueling, ultra-endurance adventure race along with the best (and craziest) in the world. Mikael has scoured the world not just in search of adventure but also in search of the fittest people on the planet.

To perform at the highest possible level, to extract that much oxygen from the air, every single part of the biologic system has to work optimally. In particular, oxygen uptake in the lungs, oxygen transport in the blood via the pumping of the heart, and oxygen extraction by the skeletal muscles all have to be maximally efficient. In many of our athletes, one of those will be superefficient—like the mountain biker we found who carries variants in the same gene that helps Andean highlanders living above ten thousand feet avoid the mountain sickness that afflicts over 10 percent of their population, or the two athletes, a male cross-country skier and a female multisport athlete, with variants in a gene that is critical for the ability of their cells to produce energy. This pathway is a major focus of research into aging and longevity, and many dollars are spent by consumers on supplements that target it. Perhaps these athletes were born with just such a "miracle drug" embedded in their genomes. We continue to search the world for these rare superhumans. In the same way Eero Mäntyranta's family helped us understand how the body makes red blood cells and provided inspiration for how we might better treat people with anemia (low red cell counts), finding the exceptional hiding in plain sight could mean harnessing the best of nature to find new treatments for those with a whole range of heart, lung, blood, and muscle diseases.

Sharlayne Tracy was a college student and mother of two living in Dallas, Texas. She felt fit and healthy, but not any more superhuman than most mothers of young kids. Her family was interested in healthy living, and her mom had enrolled in a cardiovascular health study at the University of Texas Southwestern. For this, she had undergone a series of scans that showed she was doing well and had no evidence of heart disease. When they measured her cholesterol levels, however, they were surprised, in a good way. As you'll recall from earlier, we generally hope for our "bad" cholesterol (low-density lipoprotein, or LDL) to be close to, or preferably below, 100 mg/dl. If you are unlucky enough to have had a heart attack, we recommend you try to reduce that number to 70 mg/dl or even lower if possible (though with current medications, that is not always easy). Sharlayne's mom, who was on no cholesterol medicine of any kind, had an LDL of 49 mg/dl. This was remarkable in itself but not unheard of. However, when they measured the LDL of Sharlayne herself, they had a different reaction. This thirty-two-year-old African American Texan was walking around with an LDL of just 14 mg/dl.

To understand why, we have to start in France.

French geneticist Catherine Boileau generally wears a huge smile. Her infectious energy and joyful delight in science exudes from her every pore. She was fascinated from an early stage in her career with the mathematical aspects of genetics; she enjoyed studying relationships between family members and seeing how those relationships affected their chances of carrying disease variants. But even more, she was drawn to solving mysteries of disease. One disease in particular stimulated her curiosity, a disease known as *familial hypercholesterolemia* (FH), in which patients with sky-high cholesterol levels suffer early heart attacks, some as early as their twenties. Making her crusade more urgent, it was estimated that as many as 90 percent of patients don't even know they have the disease.

By the end of the 1990s, two genes were already known to cause FH, but Boileau was bothered by the large number of affected families still without answers. She started studying families in which the two known genes had already been ruled out. By using the linkage approach discussed earlier, Boileau narrowed the cause in one family in France to chromosome 1. In this family, the first person diagnosed was a seven-year-old girl with an LDL of 236 mg/dl. Her sister's LDL was even higher at 312 mg/dl. Yet the area of the genome Boileau's team was able to identify as responsible contained forty-one genes. How to narrow that list down? Enter Nabil Seidah, an Egyptian scientist at the Montreal Clinical Research Institute, who had discovered a gene family with a rather unwieldy name. One of those genes appeared in Boileau's list of forty-one suspects. It was *proprotein convertase subtilisin/kexin type 9,* otherwise known as *PCSK9*. Working together, Boileau and Seidah established in 2003 that one specific DNA letter change at position 625 in the *PCSK9* gene was the cause of FH in the French family. A new gene for FH had been found.

The study Sharlayne's mom had enrolled in, long before her cholesterol-clearing superpower was discovered, was the Dallas Heart Study, a project run by Helen Hobbs and Jonathan Cohen. Originally from Boston, Helen speaks with the intensity and energy of a scientist driven to discover. Her main focus was the genetics of FH, and she was aware of Catherine Boileau and Nabil Seidah's discovery of *PCSK9* in 2003. Yet it was a corridor conversation with Jay Horton, a scientist who ran a lab down the hall focused on the causes of fatty liver disease, that changed her path. Horton had recently finished an experiment where he increased the amount of active PCSK9 in the livers of some of his mice. He found that this resulted in extremely high levels of circulating LDL. The finding strongly suggested that the mutation in Catherine Boileau's FH family from France wasn't disrupting PCSK9 but rather causing it to work overtime.

Think about PCSK9 molecules as being like the supervisors of a recycling agency, and the LDL receptors as the workers who actually pick up the recycling. In these FH families, these supervisors are overactive, checking the books, pulling the teams back home strictly on time so they finish work for the night having collected only half of the recycling. As a result, a lot of recycling is left on the sidewalks. Correspondingly, when normal PCSK9 pulls the cholesterol collection out of commission, lots of bad cholesterol is left circulating in the bloodstream. But what if instead these PCSK9 supervisors dozed off, leaving the collectors out all day and all night? Now every single piece of recycling is collected as soon as it is placed at the curb. That was the situation in Sharlayne Tracy, our superhuman college mom with the LDL level of 14 mg/dl. Her *PCSK9* gene was completely disrupted, leaving LDL receptors out roaming the streets all day and all night scavenging cholesterol.

The question Hobbs and Cohen wanted to address then, was, if there are people carrying mutations that inactivate *PCSK9*, would those people be protected against the effects of high cholesterol? Most things in life are easier to break than they are to enhance, and that also goes for genes; most of the time, if you mutate a gene, the change is more likely to reduce its function, so, Hobbs and Cohen reasoned, inactivating mutations shouldn't be hard to find. They decided to go look for mutations that might inactivate *PCSK9* in the participants in their Dallas Heart Study. It didn't take them long to find several genetic variants that inactivated *PCSK9* in African Americans, including in Sharlayne's mom. Sharlayne herself had inherited two inactivating variants, one from her mom and one from her dad, meaning both her copies of the *PCSK9* gene were defunct. And she was not alone; of 3,363 African American participants in the Dallas study, 2.6 percent had inactivating mutations in *PCSK9*, and these were associated with an almost 30 percent drop in LDL. What was even more remarkable, however, was that there was an almost 90 percent reduction in their risk of heart disease (inactivating variants were also found in non–African Americans though the reductions in heart disease were a little less dramatic there, perhaps because their treatment rates with traditional drugs were higher).

Once again, nature had shown the way. The competition was on: Who could develop a drug to mimic these Texan superhumans?

It soon became clear to those hoping to target *PCSK9*, however, that the usual method of finding a new drug to inhibit a protein—screening millions of small molecules to find one with the specific action desired—was falling short. This approach is usually favored because it maximizes the chance of a medicine that can be taken as a pill. It is also the approach that pharmaceutical companies are most familiar with. Yet after many years of effort, several companies realized almost simultaneously that *PCSK9* was not going to yield to this approach. The opportunity, however, was too good to miss: How else to shut it down? Drug developers turned to antibodies. These are molecules naturally created by the B cells of the immune system to neutralize attackers like viruses and bacteria. But scientists have been using them for years for their ability to recognize and stick to specific molecular targets with the precision of a lock and key. Several drug companies got to working on antibodies to neutralize *PCSK9*.

Relatively speaking, it didn't take long. In 2015, the U.S. Food and Drug Administration (FDA) and the European Medicines Agency (EMA) approved alirocumab, a drug jointly developed by Sanofi and Regeneron, as well as evolocumab from Amgen. Both drugs are formulated to be injected under the skin. The injection contains high doses of antibodies that inactivate *PCSK9* within hours.

These drugs were approved based on large, randomized, controlled trials, where neither doctor nor patient knows who got the real drug or a placebo. The first one randomized 27,564 people with known high-risk cardiovascular disease to either evolocumab or placebo. In those who received the active drug, there was a whopping 60 percent reduction in LDL from 92 mg/dl to 30 mg/dl with a 15 percent reduction in the risk of heart attacks over a two-year follow-up. Amazingly, when a subgroup of the participants whose LDL levels dropped below 10 mg/dl were examined, there were no worrying side effects. The similar study of alirocumab showed similar benefits. Each study showed about a 5 percent drop in mortality for those taking the drug.

The *PCSK9* story ignited and accelerated a decades-old interest in the use of genetics to help with drug discovery. (In the 1970s, famed drug developer and Merck CEO-to-be P. Roy Vagelos initiated a program to develop the drug finasteride based on the observation that children with genetic abnormalities associated with low testosterone also had small prostates and a resistance to male-pattern baldness.) The idea of starting with human experiments of nature, rather than with strong hunches that had to be first proved in mice, was appealing. In a classic paper summarizing the potential, Harvard geneticist and drug development pioneer Robert Plenge wrote in the journal *Nature Reviews*

*Drug Discovery* that "naturally occurring mutations in humans that affect the activity of a particular protein target or targets can be used to estimate the probable efficacy and toxicity of a drug." Drug companies were excited, too, by the possibility of a more rapid development cycle. While estimates vary, taking one new drug to regulatory approval costs between $1 billion and $3 billion once other projects' failures are built in. The sobering statistic that 90 percent of drugs that enter clinical trials fail to gain regulatory approval is a potent driver for a new approach. In the case of *PCSK9,* the relatively short period, twelve years, between the discovery of a new disease-gene association by Boileau and the approval by regulatory bodies of two new therapies was highly appealing. Commenting in *The New York Times,* Joshua Knowles, director of Stanford's familial hypercholesterolemia clinic, called it a "revolutionary approach to drug discovery."

The application of genomics to drug discovery was revolutionary for another reason as well: helping prove cause and effect. As Robert Plenge went on to say in his review, genomics could be used "to establish causal rather than reactive relationships between targets and outcomes." This second point is really key. In biomedical science, we measure many things, and we often find correlations between them. That is to say, when one thing goes up and another reliably and consistently goes up, then we say they are correlated. Correlation can also be the reverse, when one thing goes up and the other reliably goes down. It's easy to see correlations in the world: when it rains, people wear raincoats; when a house is burgled, nearby homes get alarm systems. What becomes challenging, however, is when people fall into the trap of assuming that a correlation implies that one thing *causes* the other to happen. The direction of cause and effect seems obvious to us when it relates to things we understand. We know that people wear raincoats to protect themselves from the rain; it doesn't rain because people put on raincoats. But in the realm of biological research, we often can't tell if one thing is actually causing another. Indeed, touting spurious correlations is a hobby for some people. There's even a book and a website called *Spurious Correlations* by Tyler Vigen that illustrates the ridiculousness very well. Did you know that the number of people who died falling into a swimming pool is closely related to the number of films in which Nicolas Cage appeared that same year? Or that ice cream consumption is related to shark attacks? Or that per capita cheese consumption correlates well with the number of people who died getting tangled in their bedsheets (both rose between 2000 and 2009 at exactly the same rate)? Or perhaps you could explain to me why the divorce rate in Maine dropped at the same rate as the consumption of margarine? While these examples seem ridiculous to us, that is only because we understand the world well enough to know that there could be no causal link. But what about a world we don't yet truly understand? That is the realm of science, and we often can't tell what is correlation and what is causation.

In science, proving that one thing causes another is actually quite hard, yet it is critical for showing that a drug works. Let's say we ask all our patients if they want to try a new drug. The ones who say yes, we give the new drug to. The others we follow as controls. A month later, we measure the cholesterol levels in the two groups. The group who took the drug have lower cholesterol levels. Hurrah! Case proven. Except, of course, it's not. What if the people who were more likely to sign up for the drug trial were the ones already much more interested in lowering their cholesterol and, therefore, the ones already eating a great diet and exercising every day? Indeed, it seems quite likely. This is why, to prove a drug works, we need a randomized controlled trial. A coin is tossed for everyone who elects to join the study, and they are randomly assigned to get the drug or a placebo (since sometimes just the act of taking a pill, even a sugar pill, can change people's health outcomes considerably). Randomizing everyone into two groups removes the biases that normally affect studies based on observation. Robert Califf, a Stanford and Duke professor of cardiology and former FDA commissioner, calls it "God's gift of randomization."

Interestingly, genetics allows another method of establishing one thing as a true cause of another. The approach is called, fittingly, *Mendelian randomization.* It takes advantage of the fact that the gene variants you get are essentially "randomly" allocated to you at the time of your conception. And because this happens before life in the outside world, which version of which gene you get cannot be caused by anything that happens later in your life. For example, let's say we are interested in the relationship between increased height and wealth. We find out they are correlated, but the direction of causality isn't known. Clearly, being tall could cause you to become wealthy because you are viewed as more imposing and so are more likely to succeed in that high-powered interview for that high-paying job. Or it could be that if you come from a wealthy family, your nutrition was better, so you grew to be tall. We really have no way of sorting this out from a correlation alone. But what if we have a gene that we know for sure contributes to height? Let's say, for simplicity, it is the *only* gene that contributes to height and that it doesn't do anything to impact any other pathway or behavior that might affect height or wealth (these are big assumptions, but let's make them for now). Now we can take a population of people who have either "short" or "tall" variants in this gene and divide them according to their "dose" of the height gene variants (two short genes; one short gene and one tall; or two tall genes). Whether they got short or tall variants from their parents was random chance at the time of conception and independent of any other factors in their lives that might confound the question. Now we can look in the population and see if, on average, those who were randomly assigned tall gene variants in advance are wealthier than those who were randomly assigned short gene variants. If we do this in a large enough population, and find a relationship, we can reasonably conclude that height does indeed *cause* wealth. Alternatively, if our correlation is not found when looking at height genes but only when we look at actual height and actual wealth, we can conclude other factors are at play. Note how important our assumptions are here. Nevertheless, understanding the limitations, and with the advent of large population groups being brought together with health data and genetic information (see chapter 18), we can use this technique to help us decide which pathways to target with new drugs.

A real example of how, in the absence of such techniques, drug development can go wildly wrong involves HDL, or high-density lipoprotein (the "good" cholesterol). We have known for some time that high levels of HDL are associated with fewer heart attacks. So naturally, we might imagine that a drug that increases HDL would prevent heart attacks, right? Well, not so fast.

Pfizer, one of the world's largest pharmaceutical companies, invested more money than it had invested in any previous program to develop a drug to raise HDL. It failed. Since then, every other drug aimed at raising HDL has also failed to reduce heart attacks, including those studies where HDL itself was actually infused into the patients! (One large study showed a small effect, but most believe it was because the cholesterol was shuttled from LDL to HDL: the decline in LDL was what created the benefit.) How could one of the world's savviest drug companies make such an expensive mistake?

The explanation is in the difference between correlation and causation. Heart attacks tend to happen in people with high LDL and low HDL. They also happen more often in people with gray hair. It seems obvious to us that gray hair doesn't cause the heart attack; it is simply associated with older age, which is itself the risk factor. And self-evidently, dyeing your gray hair black doesn't decrease your likelihood of a heart attack. But dyeing a patient's hair from gray to black is equivalent to what these drug companies were doing by raising the patients' HDL. How do we know?

In 2012, two researchers from Harvard University and the Broad

Institute, Sek Kathiresan and Benjamin Voigt, presented a study that used Mendelian randomization to answer exactly this question. They used both single genetic changes and a score made up of several genetic changes to test the idea that low HDL levels were like gray hair—associated with heart disease, but not causing it. What they found was a billion-dollar result. Whereas in observational studies of humans, an increase in circulating HDL was found in those with lower risk of heart attack, when the researchers concentrated on the portion of HDL determined in advance by "high" or "low" HDL genes, the association disappeared! There was no causal relationship.

To check they were on the right track, the researchers repeated the experiment with LDL and found, in contrast, a clear association between the "genetically randomized" portion of LDL levels and heart attacks. Unlike with low HDL, it was clear that a high LDL level actually caused heart attacks, meaning that targeting LDL with drugs would make sense (and indeed those drugs have saved probably millions of lives).

This finding, however, came after Pfizer and other drug companies had already spent money dyeing the "gray hair" of HDL black. Imagine now a world where instead of finding this out *after* the fact, you can start your drug discovery program with this knowledge already in hand. Not surprisingly, this is, in fact, the direction that most of the major pharmaceutical companies have moved. From now on, very few drugs will be developed without the mechanism being checked first with human genetics.

The ten-year-old from northern Pakistan was well known to the locals and to the local medical services. He would regularly perform street theater, where he would appear to harm himself with dangerous objects, but it looked so real. The reason was simple. This was not the magic trick, sleight of hand performed by sword swallowers and fire breathers. This boy would stab actual knives into his arms and walk on real burning coals. The resulting injuries were real, yet the excruciating pain that would accompany these acts in any normal person was absent. He had what is known as *congenital insensitivity to pain.*

"People assume that feeling no pain is this incredible thing and it almost makes you superhuman," Stefan Betz, a man who also suffered congenital insensitivity to pain, told the BBC correspondent David Cox in 2017. In fact, it's the exact opposite. "We would love to know what pain means and what it feels like to be in pain. Without it, your life is full of challenges." This is because pain is actually a protective mechanism that evolved to protect us from injury. Unnoticed and untreated, wounds don't heal, they get infected, and infections spread. Also, people take risks that they don't realize could harm them. The tragedy of the young Pakistani boy is testament to this; he died before his fourteenth birthday after jumping off a house roof. He believed he was invincible.

Out of this tragedy, however, came a startling insight. In 2006, James Cox, Frank Reimann, and Geoffrey Woods from the Departments of Medical Genetics and Clinical Biochemistry at Addenbrooke's Hospital in Cambridge, UK, reported a series of four families, including the one from which the Pakistani boy came, where genetic variants had reduced sensitivity to pain. Each of the affected individuals could feel warm and cold, pressure and position, but not pain. The team used a linkage/mapping strategy to narrow the genome area responsible to one containing fifty genes. The most likely gene among those fifty was one called *SCN9A,* a sodium channel gene already known to be involved in pain. Work from Steve Waxman's laboratory at Yale, and a group from the Chinese National Human Genome Center, had previously found variants in this gene that made the sodium channel overactive, thereby causing a rare syndrome characterized by recurrent episodes of severe pain. In contrast, in the UK study of the families from Pakistan, the affected individuals harbored mutations that caused both copies of this gene to be inactivated.

Here, they realized, was a potential lead on a brand-new way of controlling pain. Despite the advances of modern medicine, we remain remarkably limited in our ability to manipulate pain responses. The central therapeutic target of today's painkillers is the opioid receptor system, with medicines like morphine hitting the same system as the drug heroin. As anyone who has ever taken one of these medications knows, they are mind-altering, intensely addictive, and create a sense of dissociation from the pain sensation. You know the pain is there, but somehow you no longer care. Today, we find ourselves amid an international opioid addiction crisis that includes a black market flooded by synthetic opioids with high potency and unpredictable effects. Yet aside from addiction, there are other serious challenges created by targeting the opioid receptor. Side effects include nausea and, with prolonged use, crippling constipation. There is also the possibility of death by overdose, caused by a complete cessation of breathing. The National Institute on Drug Abuse estimates that there are 130 deaths from opioid overdose every day in the United States alone.

So can we learn something from the genetics of these families, born insensitive to pain, that will lead to a new class of painkillers without opioids' dangerous side effects? The pharmaceutical industry is working on it. Following the original publications identifying the *SCN9A* gene, there was a flurry of activity as companies tried to mimic the genetic effects of shutting down the sodium channel using drugs. The genetic effect was "clean," in the sense that affected individuals do not experience pain but also appear to have no other direct negative health consequences. Targeting the *SCN9A* protein could therefore be a magic bullet for pain. However, drug developers soon learned how hard it was to target just that one sodium channel. The reason was that there are nine members of this particular protein channel family, all of them very similar to one another. So a drug targeting one tends to bind to all of them, causing unwanted effects in brain cells, like epilepsy. Targeting just the channel encoded by *SCN9A* to mimic the families' syndrome was challenging. As a result, pharmaceutical companies are now looking beyond their usual searching ground for new drugs (their huge libraries of preexisting molecules). A promising approach is to use antibodies like those that worked so well for *PCSK9*. Another is to turn to nature; Amgen is now looking for molecules that target *SCN9A* in an unusual place: tarantula venom. Scouring the animal kingdom for superhuman biology is another way to identify new drug targets and even actual drugs. Nature has evolved a spectacular array of mechanisms for targeting the nervous systems of other animals with great precision—pain and paralysis, as inflicted by venomous spiders and snakes, are potent forces for defense or attack. Perhaps now we can use those defenses for our own precision attack on pain.

Superhumans live among us, and the genome can help us learn what makes them special. Can we harness that understanding to help the rest of us become just a little bit more special? I think we can. Yet the genome can do more, by demonstrating in advance that the mechanisms we are targeting with our new drugs are truly the ones that cause the disease we are hoping to impact. These analyses are possible today. New drugs based on the insights described here are less than a decade away. Disease should be on the retreat. The superhumans are coming.

# 18
# Precision Medicine

*Concision in style, precision in thought, decision in life.*

—VICTOR HUGO

*All of us, all of us, all of us trying to save our immortal souls,*
*some ways seemingly more round about and mysterious than others.*
*We are having a good time here. But hope all will be revealed soon.*

—RAYMOND CARVER, *All of Us: The Collected Poems*

Eric Dishman didn't know what to think. Two doctors were arguing in front of him. Clearly, nothing was clear. Whatever they had found in his tests couldn't be good, but equally, they didn't seem very sure just what it was they had found. Eventually, they turned to him. "Your definitive diagnosis is *either*"—Eric imperceptibly rolled his eyes at the idea of a "definitive" either-or, but he was too anxious to focus on that right now—"a very rare form of an adult kidney cancer *or* a very rare form of a childhood kidney tumor."

Eric was nineteen. It was 1987, and he was an undergraduate at the University of North Carolina at Chapel Hill. He had been well his whole life. But recently, he had been feeling light-headed and dizzy, and unusually for a freshman in college, it didn't relate to overindulging in alcohol the night before. He had undergone some basic tests at the student health clinic, and before he knew it, here he was, sitting in a sterile, beige room in the UNC hospital, with two kidney specialists arguing in front of him about what terrible disease he may or may not have. Nothing about the idea of cancer or a tumor sounded good, but since they didn't know for sure, how bad could it be? And though his mind was racing, the events that followed seemed to happen in slow motion, or at least that was how he would recall them in the years to come. "Because of this cancer, you aren't a kidney transplant candidate, unfortunately," they told him. Eric sensed there was more. "In fact, we think you probably have about nine months to live."

His world stopped. Few nineteen-year-olds spend any time reflecting on their mortality, but to be delivered a death sentence when you feel in overall good health, after just a few weeks of mild symptoms? Unreal. Nine months to live? He wouldn't see twenty? Where did they get such a ridiculously precise number?

And so, then came the chemotherapy. The oncologists didn't actually know what kind of cancer they were treating, so they didn't hold back. Cancer chemotherapy is based on killing the cells in your body that divide rapidly—all of them. That's why patients lose weight and lose their hair. Eric was only 120 pounds when he started. Losing 25 pounds from his five-foot-ten frame left him weak and emaciated. Was it worth spending his last few months feeling like this?

One day, after suffering with abdominal pain, he was going for an endoscopy procedure to check for stress- or chemotherapy-induced ulcers in his stomach. For this kind of procedure, they used minimal sedation, and so he brought along a friend for moral support. On the way, she asked how much research he had done on this mystery cancer the doctors were treating. Eric realized the answer was, "Not very much." His friend persuaded him to ditch the endoscopy and instead marched him directly to the Duke medical library. There, together, they pored through dusty medical journals and learned what was known about renal cell carcinoma, the form of cancer his specialists thought he might have. Startlingly, the only studies they could find about renal cell carcinoma were in patients decades older. No matter how hard they looked, they couldn't find a study describing this cancer in younger patients. How could these patients and their prognoses possibly be relevant for him, they wondered? He was nineteen. In contrast to the equivocations of the doctors who diagnosed him, his friend's advice was clearly stated: "They don't know anything about you! Go live your life!"

He did, never dreaming that this condition and this decision would bring him face-to-face with the cutting edge of precision medicine.

By the time I met Eric in 2015, he was a seasoned professional in the prime of his career. He had spent the past several years of his life at Intel as vice president of the Health and Life Sciences group, thinking up ways technology could improve the health care experience and, specifically, make it more patient-centered. As we sat having breakfast at Peet's Coffee in the Clark building at Stanford one morning in the summer of 2018 (he was giving a talk a couple of hours later at our annual Big Data in Biomedicine conference), he pulled out a handful of pill bottles and plopped them on the table. "Burdens of being a transplant patient," he said with a rueful smile as he swallowed a handful of pills. "At least I'm alive." He laughed. Later, in his talk, he added a more celebratory note: "I just turned fifty! I mean, it's a miracle I turned fifty. I was never supposed to turn twenty, thirty, or forty."

In person, Eric is warm, gracious, with an infectious laugh and a relaxed conversational style. Yet you can feel, just below the surface, a fierce sense of urgency to make a difference, to really use this time he's been given to give back. He is prone to phrases like *from an "impatient" patient perspective.*

As it happened, the doctors never really worked out what type of tumor Eric had. After the conversation in the Duke library that day with his friend, he changed his approach. He would take all the chemotherapy, but he would live every day to the fullest. He no longer listened when doctors told him his time was limited. What did they know?

After ten years of chemotherapy and being told pretty much every year that he wouldn't live another year, he was also ready to dictate the terms of his acquiescence. He went to his doctors and asked if, from now on, they would tailor the chemotherapy regimens to his love of the mountains and especially his love of winter sports. Since electronic health records don't generally have a field for patient goals, he asked them to put his preferences in the "Nickname" field. From then on, he wanted his record to read, "Eric Dishman (optimize for snow time)." And if that didn't fit, he requested, "Eric Dishman (Snow Dude)." He had a point. So much of medicine is organized around the existing culture, history, and traditions of doctors. It is not focused on the beliefs, goals, and desires of the patient. By asking his doctors to optimize their choice of chemotherapy to maximize his time on the mountains (and even specifying his preference for nausea over headache as a side effect), he bought himself back some control of his narrative. He became more than the next patient in line. That experience was to shape his career.

Originally trained as a social scientist, with an interest in anthropology and ethnography, he was well attuned to humans and how they organize (or don't organize) themselves. He became familiar with the qualitative research methodologies common in sociology and psychology before entering the world of tech in the early 1990s as the first intern for Microsoft cofounder Paul Allen's think tank, Interval Research. When he later moved to Intel Corporation, he got to know another computer age pioneer: Andy Grove, Intel's CEO and a legendary business leader. Grove was, in his later life, diagnosed with prostate cancer and Parkinson's disease—and devoted a significant portion of his fortune, as well as his influence at Intel, toward improving health care. He and Eric shared the same passion for putting the patient at the center of their own experience: "Own your own health care data, like your life depends on it," Grove was fond of saying.

As a VP at Intel, Eric was looking for opportunities to bring twenty-first-century Silicon Valley tech to what at times felt like eighteenth-century medicine. Genomics appeared to be a good fit. The computer programs that were used to crunch through genomic data were often written by professors and their students (like me and mine), not by professional software engineers like those at Intel. By enlisting a team of professional software engineers, Intel's Health and Life Sciences group was able to greatly speed up the open-source software used for interpreting genome sequences. This work exposed him to many start-up companies in the genome sequencing space. In 2011, as his kidneys failed and he tried to avoid dialysis so he could continue his cancer treatment, a contact at one of those companies offered to sequence his tumor. After all, the doctors had never really managed to classify the tumor effectively or to identify the gene mutations fueling its growth. Maybe this company could give it a try?

By this time, genomics had begun to impact the world of cancer in revolutionary ways. Early genetics studies from the 1990s had focused on specific mutations that patients were born with that predisposed them to cancer later in life (mutations in the *BRCA1* and *BRCA2* genes are good examples). By the early 2000s, however, focus was shifting to new mutations that arose in the cancer cells themselves that turned up the throttle on unrestricted growth. As a result, oncologists were beginning to understand and classify cancer not according to the tissue in which it first occurred (the traditional way of classifying cancer—as "lung" or "breast" cancer) but, rather, according to the unique biology of the tumor. Certain genetic causes of unrestricted tumor growth tend to recur, they found, and these might be good places to focus therapy. Sometimes, the best treatment might not be the one usually used to treat a cancer arising from that organ or tissue but rather one focused on the cell growth pathway that was disturbed.

In Eric's case, a genetic comparison between his normal cells and his kidney tumor cells revealed cell growth pathway perturbations more common to pancreatic tumors. This was probably why his kidney doctors had been confused for all these years. But that also meant there was a chance he could benefit from chemotherapy normally reserved for pancreatic tumors. The idea of tuning chemotherapy to the underlying genetics of the tumor was new, and he was the first patient his oncologists had worked with where they had access to full genome sequencing both of the patient and the tumor. But Eric was no ordinary patient. They agreed to try the pancreatic cancer chemotherapy. Within a short time, Eric went into full remission. The personalized approach worked beautifully, like no prior treatment had. After a period of being deemed cancer-free, he then proved his original doctors wrong again by becoming a transplant candidate. At forty-four years old, a full twenty-five years after being given nine months to live, he received a new kidney and a new life. During a 2013 TED Talk about his experience, Eric brought his kidney donor up on stage. The audience stood for an ovation that celebrated the courage it takes to survive decades of treatment for a mystery cancer, and the selfless altruism manifest in donating a kidney to a complete stranger.

As Eric and I sat in that coffee shop in 2018, our conversation turned to planning his post-Intel future. Having survived 57 rounds of cancer treatment, 1 kidney transplant, and 371 different medications, he had decided it was time for a new role—one positioned at the very heart of our nation's precision medicine efforts. It was time to help champion his cause at the highest levels of the U.S. government.

Right from the start, it seemed obvious to Francis Collins—head of the National Human Genome Research Institute, leader of the Human Genome Project—that "we were never going to be satisfied with one genome." Finding a way to sequence and compare genomes across whole populations was clearly going to be the next big mission. In 2003, soon after the completion of the Human Genome Project, Collins gathered together critical thinkers: among them was Teri Manolio, a physician and epidemiologist with a background in large-scale population studies; Eric Boerwinkle, a straight-talking geneticist from Texas who had run pivotal genetics studies in high blood pressure and diabetes; Wylie Burke, a University of Washington professor specializing in the ethical and policy implications of genetic studies; and David Altshuler, a rising star in the newly formed Broad Institute for genetics research in Boston. The team understood that characterizing human genetic variation across hundreds of thousands or even millions of people was going to be the key to realizing the potential of the Human Genome Project. Only by analyzing large populations could the genetic variants that really mattered be distinguished—which variants caused disease, which were benign.

Bold initiatives were already starting to emerge elsewhere in the world. The forerunner was deCODE Genetics. Based in Reykjavík, Iceland, it was founded in 1996 as a public-private partnership to which most of the country's three hundred thousand people eventually signed over health records, genealogy records, and a DNA sample. Iceland is a particularly rich resource for such a study. Settled by Viking explorers from Norway and the British Isles in the ninth century A.D., the population has remained relatively isolated, from a genetic standpoint, since. Indeed, most of the population can trace their lineage back to just a few common ancestors. Against this relatively uniform genetic background, rare variation stands out, offering great power for the discovery of human genetic variants predisposing to disease.

The potential for scientific discovery in Iceland did not go unnoticed by Kári Stefánsson, an Icelandic neurologist and geneticist at Harvard. In the late 1990s, he became interested in the potential of his native Icelandic community for the study of the genetics of multiple sclerosis. After struggling to obtain support from the U.S. National Institutes of Health, he decided instead to raise private capital and returned to his homeland with a grand vision for studying the genetic basis of a range of human diseases.

Shortly thereafter, the United Kingdom launched its own ambitious effort. In the late 1990s, it became clear to leaders in Britain's academic medical and biotechnology sectors that, thanks to its National Health Service, the UK was uniquely positioned to take advantage of the genomics revolution. With its sixty million residents enrolled in one health care system, it would be feasible to connect genetic information to detailed health record information across the entire population. This would create an enormous opportunity to better define diseases and find new drug targets to defeat them. Two individuals, in particular, laid out a compelling vision for this new kind of medicine. In 1998, John Bell, a Canadian-born immunologist and geneticist and future Regius Professor of Medicine at Oxford University, wrote a remarkably prescient piece for *The British Medical Journal* extolling the benefits of—someday—using genetic insights to precisely tailor therapies to the underlying mechanisms of disease, and to predict disease risks in those who are not yet ill. The following year, George Poste, chief science and technology officer for the British pharmaceutical company SmithKline Beecham, put this vision in the context of the UK's National Health Service. Writing in the journal *Science,* he and colleague Robin Fears described a biobank that would link genetic information with lifelong health data. The vision took shape at a key meeting in Bell's office in 1999, where epidemiologists Richard Peto, Valerie Beral, Rory Collins, and Christopher Murray came together to discuss what such a study could look like. Stimulated by these ideas, the Wellcome Trust, along with the UK government's Medical Research Council, committed to the project in mid-1999. The vision was bold: recruit five hundred thousand Britons to donate their medical data, blood, and DNA and share it with any qualified researcher worldwide. Rigorous debate followed in public forums and private meetings, with geneticists arguing that genetic insight would be maximized by comparing patients with specific diseases, and epidemiologists arguing that the study was about much more than genetics. In the end, it was the leaders of the funding bodies, George Radda at the Medical Research Council and Mark Walport at the Wellcome Trust, who pushed it over the line, leading to the announcement, on April 29, 2002, of joint funding of 45 million GBP ($56 million). UK prime minister Tony Blair highlighted the shared vision at the Royal Society. "We have a unique resource in this regard in the national health service," he said. "There are crucial issues of privacy of genetic information that we need to deal with. But our national, public system will enable us to gather the comprehensive data necessary to predict the likelihood of various diseases—and then make choices to help prevent them." In 2003, the first chief executive of the UK Biobank was appointed, and in July of that year, a science committee was formed, chaired by John Bell. In September 2005, current principal investigator and chief executive Rory Collins was appointed.

It was against this backdrop that Francis Collins and his team of forward-thinking geneticists laid their plans. The time was right, they believed, for the United States to consider a national project of its own—one that considered the country's unique lifestyle factors, environmental risks, and enormous ethnic diversity. Ideally, this would involve sequencing the genomes of hundreds of thousands or even millions of people to glean insights into health and disease. Yet the costs of sequencing that many people remained prohibitive. Gene chips, however, on which hundreds of thousands of preselected genetic variants could be measured, were dropping in price. That data could be obtained for a few hundred dollars per person. Although not as deep a characterization as sequencing could provide, Collins viewed it as an affordable way to correlate genetic variation with disease across a large population.

The group put a budget together, and Collins headed to Capitol Hill to speak to Bill Frist, the Senate majority leader and a former heart surgeon. Frist was supportive of the idea, but the size of the ask was just too great for a Republican administration focused on tax cuts and lowering spending. At an impasse, Collins wrote up an opinion piece for the journal *Nature,* describing his vision of a large-scale study of hundreds of thousands of individuals: one that would analyze each participant's genes and also the health issues they developed over time. His hope was to try to stimulate momentum and broader discussion. When the article appeared in May 2004, it landed, he remembers, with a dull thud. "Scientists thought I was naïve," he told me. "They said that as a geneticist, 'Collins doesn't know epidemiology' [the study of populations]. They said I couldn't do math." The implication was that it was too much money. The project was put in cold storage. Indefinitely.

Meanwhile, in the mid-2000s, Barack Obama, the junior senator from Illinois, had also been thinking about health. He had not grown up in a family with any close links to the health care profession. Fortunately, he had mostly avoided doctors as well, with the exception of a run-in with a barbed wire fence he suffered while mud-sliding in Indonesia, which left him with twenty not-very-expertly-placed stitches from elbow to wrist. But at the urging of his health and education advisor, Dora Hughes, the senator had learned about the nascent field of personalized medicine. Hughes had previously worked with a young genetic counselor named Jennifer Leib on the Senate Committee on Health, Education, Labor, and Pensions. Leib had trained at Johns Hopkins, had an interest in public policy, and, at the time, was working in government affairs for Affymetrix, the competitor to Illumina in the gene chip market. Like Francis Collins, Leib recognized that these inexpensive chips (which are still used today by direct-to-consumer companies such as Ancestry and 23andMe) were clearly a path to the personalized medicine era. They could put genetic analysis within reach of almost everyone, generating insights that could then be used to improve health care. Leib and Hughes talked about what legislation might look like to help achieve those aims, and Leib gave Hughes one of the chips to show the senator. It was rectangular, about two inches by one inch and a quarter of an inch thick. If you looked carefully in the middle, you could see a square waffle pattern where the business of detecting DNA variants would occur. Hughes brought the chip to her next meeting with the senator. When she explained that these chips were about to revolutionize genetics, Obama was intrigued. He was a science geek, after all, and loved technology. He held it up to the light, wondering if he could "see" the DNA. (Hughes explained that the molecules were too small.) But as she described what these chips could do, warming to her theme of an impending personalized medicine revolution, and its potential to help patients like his mother, who had been diagnosed with ovarian cancer, Obama was hooked. He gave the green light for exploration of potential legislation to advance the field. Hughes, Leib, and another young staffer named Eduardo Ramos got to work. A central feature of the bill that was eventually introduced to committee (S.976: The Genomics and Personalized Medicine Act of 2007) was helping different parts of the government work together to accelerate genomic medicine. The bill also aimed to establish a national biobank for genomic and health record data. There was an emphasis on the education of health care providers. It was bold, and it was ahead of its time. Obama, along with Richard Burr, a Republican senator from North Carolina, introduced it into the Senate. Their proposals could have had a transformative effect on medicine in the United States. But with a divided government, the bill never saw the light of day.

Fast-forward to 2009. These futuristic visions began to converge when Obama, now president, appointed Francis Collins as director of the National Institutes of Health. Collins found Obama to be "incredibly interested in medical research and medical opportunities." He described one-on-one sessions he would have with Obama in the Oval Office and how impressed he was, not only that the president could be completely zeroed in, despite the pressures of his job, but also that he would want to learn, to be taught something. "What a sharp mind. And what an ability to focus," he remarked to me. Not surprisingly, Obama looked toward his own White House Office of Science and Technology Policy for guidance about what to tackle first. Established in 1976 by Congress, the office held a broad mandate to advise the president on science and technology. Under Obama, its director was John Holdren, an MIT- and Stanford-educated physicist.

John Holdren had a strong family history of breast and ovarian cancer, and his daughter Jill worried about risk that could be transmitted to future generations through him. Frustratingly, because she herself didn't have a first-degree relative who was affected, DNA testing for her was not covered by insurance. Her dad was eligible for testing, and several years after he was appointed as head of Obama's Office of Science and Technology Policy, at the urging of his daughter, he was tested and found to carry a pathogenic mutation in the *BRCA1* gene, signaling a high familial risk for breast and ovarian cancer. Jill was quickly tested for this family mutation and, having been found to carry it, immediately underwent preventive surgery to remove her ovaries, uterus, and breasts. But it was too late: the doctors found cancer in her ovaries that had already spread to the nearby tissue. At operation, the surgeon removed the cancerous tissue, and a time of anxious waiting began.

Jill Holdren had a background in public health. She had met the president and his family at a holiday dinner, and as she approached these surgeries, she received encouraging notes and gifts from Obama. She also spent this time diving deeply into research on the genetic predisposition to cancer and exploring broader issues at the heart of personalized medicine. She soon came to believe that democratizing access to genetic testing was key to preventing disease. She hoped that her father could share some of these thoughts with the president. Going one better than that, her father arranged for her to talk to the president directly. So in May 2014, Jill met President Obama in the Oval Office, where, after the usual pleasantries and pictures, she launched into a brief discussion of what she had learned about cancer, genomics, the role of large-scale patient data, and how these approaches might inform the future of health care. She asked Obama about his own family history of ovarian cancer, and they discussed for some time his own mother's diagnosis. Jill suggested Obama should himself be tested (his sister Maya Soetoro-Ng revealed in an advocacy video sometime later that she and her brother both tested negative). Obama later mentioned to John Holdren how profoundly affected he was by this conversation and remarked that rarely had he heard this case for the revolutionary impact of genomics on medicine made with such clarity and detail.

Two days later, John Holdren left for Dulles International Airport to catch a plane to the UK when the president called to say that he had been thinking about his conversation with Jill and that he wanted Holdren's team at the Office of Science and Technology Policy to design an initiative to exploit the potential that she had outlined. It was a Thursday, and he wanted a draft on his desk the following Tuesday. From the car, Holdren called his associate director for science, Jo Handelsman, and also Eric Lander, the cochair of the President's Council of Advisors on Science and Technology, and Marjory Blumenthal, an expert in systems engineering policy in health care. He asked them to work together on a draft over the weekend that he would look at on Monday after his return from the UK. They did so, adding to the team another genomics and science advisor, Tania Simoncelli.

The broader team worked the draft for two days and got it to the president a day late. A week later, in late June 2014, Obama gathered a group in the Oval Office to discuss the proposal they had drafted, adding NIH director Francis Collins and FDA administrator Peggy Hamburg.

Brilliant minds came together in the most famous elliptical room in the world that day, to bring to life an idea that had simmered for a decade: a large-scale genomics initiative aimed directly at improving human health. Details were coming into focus: they would try to recruit a million Americans to take part in a research study, administer them comprehensive questionnaires, ask them to share their health record data, and undergo genome sequencing. It would be the largest study of its kind. It would prioritize diversity in recruitment (too many genetics studies had been performed in populations of only northern European origin, they felt). One more instruction came directly from the president, and it made Collins and the others swallow hard.

"I will not support a program," the president said, "where you are asking people to be your partners, but you're withholding data from them." Instead, Obama insisted that all participants should have access to their genetic data and, indeed, any data gathered about them for the study. For context, the norm in genetic research at that time was to analyze the data without linking it to individual patients' names or other identifying details. Even the researchers had no idea whose data belonged to whom. To set up a system for returning data to the participants? To treat them as partners? No prior project came close to the kind of participant partnership and transparency Obama was suggesting.

As 2014 turned to early 2015, it appeared that this bold idea was about to gain a national platform worthy of its audacity. Rumors were swirling that precision medicine might be mentioned in the State of the Union address. I was driving home in my car on January 20, 2015, when I heard on my radio the forty-fourth president of the United States begin his sixth State of the Union. Once I got home, I immediately turned on the TV.

Obama shook hands on both sides of the aisle as he entered the joint session of the House of Representatives that night and worked his way to the stage. He looked up to greet Vice President Joe Biden and Speaker John Boehner, who sat behind him, then turned to the lectern, the assembled representatives, and the American people. He described the end of the combat mission in Afghanistan. He described the economic revival. Then, twenty-eight minutes into his speech, as he hailed the prowess of American science and technology, he warmed to a new theme: "I want the country that eliminated polio and mapped the human genome to lead a new era of medicine—one that delivers the right treatment at the right time." He continued, "So tonight, I'm launching a new Precision Medicine Initiative to bring us closer to curing diseases like cancer and diabetes. And to give all of us access to the kind of personalized information we need to keep ourselves and our families healthier." Then, in an ad lib that at once echoed the depth of his enthusiasm and the room's rising applause, he added, "We can do this!" In a polarized political world, it was a punch-the-air moment that was, notably, the only line in the entire speech that brought both sides of the aisle to their feet.

Twenty-eight years after Eric Dishman received a death sentence for kidney cancer, twelve years after Francis Collins's *Nature* paper describing the virtues of a large population study, eight years after a personalized medicine bill from Senator Barack Obama failed to make it out of committee, and a mere eight months after Jill Holdren described to him her personal story of early intervention for *BRCA1* associated cancer, the official era of precision medicine had begun.

A frantic ten days later, during which time staff members worked round the clock to flesh out details, in the East Room of the White House, Obama expanded on the announcement in a televised address. Francis Collins introduced the president flanked by a red, white, and blue model of DNA that had been commandeered at the last minute by White House staff from the office of the National Human Genome Research Institute director, Eric Green. In his remarks, Obama outlined the four main tenets of the new plan: to work with the National Cancer Institute to advance individualized cancer care; to work with the FDA to ensure a path through regulatory hurdles for new genetic tests; to work with the NIH to build a diverse cohort of one million people willing to share their health data and genomic data for the greater good; and, finally, to ensure through all of this the protection of genetic privacy.

In the East Room that day, Obama was joined by invited guest and cystic fibrosis patient Bill Elder, a twenty-seven-year-old medical student who had been diagnosed at the age of seven. Bill was one of the first patients to get a new drug, ivacaftor, fast-tracked by the FDA for its positive effects in a targeted subgroup of patients with cystic fibrosis, a great example of precision medicine in action. Obama announced that he would include $215 million in his budget as an initial investment toward the goals of the Precision Medicine Initiative. More substantial funding was earmarked in, and later provided by, the 21st Century Cures Act.

Over the final two years of his presidency, Obama's commitment to this effort and his love of science and technology became even more apparent to those beyond his inner circle. At Stanford, we were lucky enough to work with several agencies on aspects of the Precision Medicine Initiative, and I heard stories of the president clearing his schedule so he could hear about its progress. His personal investment helped the initiative to move forward at lightning speed.

At the FDA, there was a major focus on removing barriers to clinical genetic testing. A new project called precisionFDA arose to facilitate early collaboration between companies developing exciting new technology and the regulatory body that would eventually need to approve it. The precisionFDA project included a shared cloud computing environment where researchers and companies could test their new tools for analyzing genomes. We hosted a hackathon at Stanford to help kick-start the project, and my graduate student Rachel Goldfeder put up the first tool.

The White House team took a major role in coordinating the initiative, most notably Stephanie Devaney from the Office of the Chief of Staff, and Claudia Williams, the senior advisor for health technology and innovation. Devaney, with a Ph.D. in molecular genetics from George Washington University, led the coordination of the effort with federal partners. Williams brought twenty years of health policy experience to her role and recognized the unique power of the White

House to bring key partners together to advance the agenda. I got to see this firsthand when we hosted a roundtable with Williams at Stanford and saw her skillfully lead discussions between thought leaders from health care, industry, and academia. We also worked closely with Williams and Devaney for an event at Carnegie Mellon University in Pittsburgh at which Atul Gawande interviewed President Obama about his love of science and technology before touring around various exhibits including cybernetics, Mars exploration, drones, autonomous cars, and our own Stanford exhibit, which demonstrated the power of digital tools like smartwatches in more precisely measuring health. In a particularly memorable moment, two of my graduate students, Anna Shcherbina and Jessica Torres, met not only Francis Collins but also the U.S. chief data scientist, D. J. Patil, and Megan Smith, the U.S. chief technology officer.

Meanwhile, momentum was gathering behind the one-million-person study. A series of advisory sessions was held around the country, chaired by the NIH deputy director for science, outreach, and policy, Kathy Hudson. At the Silicon Valley meeting, Francis Collins himself attended. He had begun to reflect on what sort of person could help really push forward this large cohort study. It would need to be someone who was a proven leader; someone who understood research and technology; someone who really understood the patient and participant viewpoint. They would not need to be someone with significant NIH experience. In fact, someone from the "outside" might even be preferable. As the room quieted that day for the start of the meeting, the host stepped up to the microphone to welcome everyone to his company headquarters in Santa Clara, California. "Welcome to Intel!" Eric Dishman began.

"He turned me down several times," Collins later told me as he described his pursuit of the "Snow Dude" given nine months to live at age nineteen. But the NIH director didn't get where he is today without persistence. Eric Dishman was announced as the director of the Precision Medicine Initiative Cohort Program in April 2016. Shortly afterward, it became known as All of Us, reflecting the mission articulated in those exact words by the president in his State of the Union address: that participants should be partners in a program reflecting the full diversity of the U.S. population. The president seemed particularly pleased that a patient/technologist was leading the way and was moved by Eric's personal story, which had been relayed to him by his wife, Michelle, after she saw Eric's TED Talk.

Along with Eric and the NIH team, several key groups around the country were brought in to help manage All of Us. At Vanderbilt University, Josh Denny, a physician and informatics specialist, was appointed to lead the computer infrastructure team. At Scripps in San Diego, Eric Topol was selected to coordinate multiple aspects, including the digital technology like the smartphone app and the wearables component. Eric is one of the most celebrated physicians in the world—former chief of cardiology at the Cleveland Clinic, bestselling author, futurist, physician, and scientist (and with a Twitter account that should perhaps be sanctioned as a national monument). In 2020, the leadership team adopted new roles with Josh Denny taking over as chief executive officer, Stephanie Devaney taking up the reins as chief operating officer, and Eric Dishman taking on a new role as chief innovation officer. Together, the project's leaders and their teams of collaborators have recruited hundreds of thousands of participants in record time, beating their own targets for ethnic diversity. They are on track to reach a million participants by 2023.

It is a trailblazing and unusual project within the NIH—and not only because its vision came directly from the desk of the president of the United States. Participants enroll and provide digital consent via a mobile application—a convenience almost unheard of in medical studies—and NIH promises to return information about health and genetic risks to participants. The seeds of a new approach, a partnership between researchers and the public, so aptly articulated and championed by President Obama, are beginning to take root.

Meanwhile, the other national initiatives that first inspired All of Us have borne fruit. Consider deCODE in Iceland. With passion, strongly held opinions, a contempt for accepted wisdom, and sheer force of will, founder Kári Stefánsson assembled and led a population genetics discovery engine the likes of which the world had never seen. Following the company's launch in 1996, studies from deCODE dominated the pages of *Nature Genetics*, one of the leading genetics journals, over the following two decades as the company first characterized Iceland's population with gene chips, then later full genome sequencing. Seminal discoveries of genetic predisposition across dozens of common diseases, including cardiovascular disease, cancer, and psychiatric disease, have been reported along with fundamental insights into human genetic diversity and the effects of parental age on the frequency of mutations arising anew in children.

Along the way, deCODE assembled an enormous online database known as the Íslendingabók (Book of Icelanders), detailing the family relationships between every Icelander throughout the nation's history. In the smartphone era, this readily available data has some unforeseen benefits. For example, young people can take advantage of a dating app called ÍslendingaApp, designed to prevent accidental pairings of related individuals. Prior to asking someone on a date, a young person can type in the name of their intended and find out how closely they are related, so they can choose a more compatible partner if necessary. In fact, a couple actually on a date can even bump their phones together and find out the answer immediately. This led to the slogan "Bump the app before you bump in bed," demonstrating not only Icelandic engineering ingenuity but a sharp—possibly genetically encoded—Icelandic humor.

A look to the UK Biobank, however, provides insight into the sheer impact that All of Us may someday have. Today, the UK Biobank stands as an exemplar—a notice to the world of what voluntary sharing of data, coordinated collaboration, and visionary leadership can achieve. From the beginning, the UK stakeholders pledged to make the data (health records and genetic information) freely accessible to any qualified researcher. True to their word, the first high-profile paper to emerge from the data set was published in 2015, not by a British researcher but by Swedish physician and epidemiologist Erik Ingelsson. In addition, that year, the sharing of gene chip data on participants made possible, overnight, the type of large-scale studies that would have taken groups of investigators many years and many millions of dollars to generate just a few months before. Suddenly, these studies were only a few lines of code away, to any qualified researcher worldwide. My colleague Manuel Rivas at Stanford has even built an online engine for anyone who wants to explore a disease or a gene or even a favorite region of the genome. It is like Google Earth for the genome and is available to anyone in the world who can access the internet.

The impact of the decision to make the data broadly available, and the selflessness of the participants in contributing their data, cannot be overstated. At the time of this writing, there have been more than one thousand scientific papers published, each meeting the criterion of a focus on any health-related topic that is in the public interest. Seminal new findings in cardiovascular disease, cancer, diabetes, sleep, asthma, psychiatric disease, aging, and retinal disease have been reported, including many methodological advancements in machine learning, genetics, and the statistical determination of cause and effect. Diet and exercise studies are a common focus. One recent study even examined the British proclivity toward tea, coffee, and alcohol, connecting participants' intake of those drinks with variants in genes for bitter taste sensation.

At Stanford, we have used the data in many different studies ourselves. James Priest, for example, the pediatric cardiologist you met in chapter 12, has a particular interest in why some babies are born with physical heart defects. One of the most common is a bicuspid aortic valve. The aortic valve, which lets blood out of the heart, normally has three flexible flaps, or cusps, that open and close to control the flow of blood. Some individuals, however, are born with an unusually shaped valve with only two flaps instead of the usual three, which can put them at risk for narrowing and leaking of that valve later in life.

As it turns out, the UK Biobank has obtained full-body MRI scans of more than a hundred thousand individuals, making it one of the largest imaging studies in the world. Together with colleagues in Stanford's computer science department, we were able to use artificial intelligence techniques in four thousand scans to identify the participants with bicuspid aortic valves, then analyze their genetic data for possible causes. Our study identified new regions of the genome important for heart valve development and heart valve disease, something that had been impossible before the UK Biobank, because no prior researcher had access to such a large group of people with both heart imaging and genetic information. A recent commitment to sequencing the genomes of all the participants will elevate the UK Biobank yet further.

Perhaps the greatest compliment for the UK Biobank came from Kári Stefánsson himself, the acerbic Icelander who founded deCODE. Not someone prone to compliments or hyperbole, he said, "In my opinion, the UK Biobank is the most significant initiative ever in biomedical research."

The UK Biobank also helped fuel another major genomics program within the UK, driven by genomics leaders and by Prime Minister David Cameron, whose son Ivan was born with Ohtahara syndrome, a rare, debilitating, and progressive epileptic condition. The program, named Genomics England, committed to the sequencing of one hundred thousand patients' genomes over five years when it was announced in July 2013. Hitting that mark in 2019, the UK government has now committed to sequencing one million genomes between the UK Biobank and Genomics England with a commitment to generating at least gene-chip-level data on up to five million people, a feat that would make it the largest program of its kind in the world.

While the All of Us team asking one million Americans to contribute DNA may sound like a lot, that number is dwarfed by the number of DNA samples currently stored in the private sector. Today, globally, the largest collections of DNA samples are held by direct-to-consumer genetic testing companies. These companies, of which 23andMe and Ancestry are the most well known, offer to analyze your DNA from a mailed-in saliva sample and return genetic insight into your ancestry and certain aspects of health—all for around one hundred dollars (23andMe will even tell you what percentage Neanderthal you are). It is popular: by 2020, an estimated twenty-five million individuals worldwide are customers of these companies.

Yet while companies like Iceland's deCODE, 23andMe, and Ancestry have amassed genetic data on tens of millions of people, the data is accessible to the world's research community only through direct collaboration (in contrast to the UK biobank or the plan for All of Us). When pharmaceutical companies started to embrace genetically informed drug discovery, as described in chapter 17, deCODE was a readymade opportunity. The pharmaceutical giant Amgen purchased deCODE for $415 million in 2012, spinning out the computer system that managed the genetic data into a separate company. 23andMe, too, has recently moved toward exploiting the sheer scale of their customer base for drug discovery with a series of licensing deals and the high-profile hiring of Richard Scheller, a former executive vice president of research and early development at the biotechnology company Genentech. Meanwhile, another biotechnology company, Regeneron, in 2014 partnered with Geisinger Health in Pennsylvania to recruit more than one hundred thousand individuals willing to share their health record data and their DNA. While Regeneron has focused on combining this proprietary genetic data with disease data from the patients' medical records to accelerate drug discovery, the Geisinger team has been returning genetic results to their patients to enable proactive preventive health care.

Several other large-scale projects have also generated a vast amount of health-related data on hundreds of thousands of individuals. These studies have made data available to a narrow group of researchers. They include the China Kadoorie Biobank, funded by the Wellcome Trust to help investigate the genetic and environmental causes of common chronic diseases, and the Million Veteran Program, funded by the U.S. government. Both have been tremendously successful in recruiting and following participants, with more than five hundred thousand individuals consented in each. Consistent with the narrower data sharing in these studies, the number of research papers from these efforts is smaller, but researchers have reported insights into social determinants of health in the Chinese population and insights into the genetic causes of high cholesterol in the veteran population. This latter study, led by my Stanford colleague Tim Assimes, was a good example of the power of a more diverse population. The greater diversity in ancestry of the veteran population led to the discovery of many new genetic variants associated with disease compared to prior studies in more homogenous northern European (white) populations.

What all these studies have shown is that the larger the study, the greater the power for discovery, and the greater our confidence in those discoveries. If the data from each of these studies could be combined, that would supercharge global efforts to analyze biomedical data for the common good.

When it comes to the meticulous process of merging data gathered by a wide range of scientists, for a wide range of reasons, into a new and cohesive databank, Daniel MacArthur is without peer. MacArthur is an Australian geneticist who combines a considerable intellect with generosity and a strong line in bitingly sardonic humor. Starting in 2012, he led an effort at the Broad Institute to bring together and make publicly available the aggregated genetic data from hundreds of thousands of people. These people had all individually signed up for very different studies, ranging from the Framingham Heart Study to the Women's Health Initiative. Data came from sources as diverse as Bangladesh and Estonia, from individuals suffering from schizophrenia in Taiwan, and from patients with bipolar disorder in Sweden. All told, to date, these projects harbored sequencing data for some 140,000 individuals from a range of ancestries. As well as being a critically important resource for understanding the genetics of populations, the database, now called gnomAD ("no-mad") became an instant go-to resource for interpreting patients' genetic test results. It used to be that if we found a new genetic variant in a patient, one that hadn't previously been reported to cause disease, we would check a database of one hundred anonymous, mostly Caucasian blood donors, to determine if the variant was common in the broader population and if so, how common. This was critical information. A common variant, one found in lots of generally healthy people, could not be the cause of a rare and devastating disease. But as we soon learned, one hundred controls was woefully inadequate for determining if a gene variant is rare. Now thanks to MacArthur's team, anyone in the world can instantly access the results from many tens of thousands of individuals across multiple different ancestries to answer that question.

In short order, MacArthur's project entirely transformed our ability to distinguish disease-causing genetic variants from genetic variants that are relatively harmless. Consider one patient of mine, Theodore

Carter, who first came to medical attention in 2010 as a high school track star. He was a middle-distance runner who had collapsed after a race. In the medical workup, it was felt a recent viral infection may have explained the collapse, but his heart was also found to be mildly thickened, suggestive of hypertrophic cardiomyopathy. This thickening of the heart muscle, however, was not the dramatic kind seen in patients with a slam-dunk diagnosis (like Leilani Graham, whom we met in chapters 15 and 16). Theodore's heart, instead, showed a mild increase in thickness that existed in a gray zone. It wasn't clear if the changes could simply reflect his intense athletic training. To help clarify, his pediatrician performed a genetic test. Sure enough, Theodore had a variant in a gene known to cause hypertrophic cardiomyopathy (*MYH7*). This variant was not found in a small sample of one hundred blood donors but was found in one other patient with HCM, leading him to conclude that Theodore suffered from a mild form of hypertrophic cardiomyopathy. As a result, he suggested that the safest approach was for him to avoid competitive track. This restriction was incredibly tough for him, as, more than just a leisure-time pursuit, his sense of self was tied to his athletic career, and he was looking toward a college scholarship. Yet with the information available at the time, his *MYH7* variant looked too real to ignore.

Fast-forward a couple of years to 2012, when the first large-scale population genetic databases became publicly available. Now instead of one hundred people, we could check the genomes of thousands to decide if a variant was too common to cause a rare disease. I remember the actual day these databases were switched on. Our lead genetic counselor, Colleen Caleshu, spent an entire afternoon simply looking up the variants we had found in patients over the previous five years. The dynamic range of her facial expressions hit an all-time high that day. Sometimes, the results confirmed our earlier conclusion: the variant was absent from thousands of people (a much more convincing sign of rarity than not being found among one hundred people) and therefore highly suspicious to be the cause of a rare disease. For Theodore's variant, however, the news was quite different. His variant had been found in many individuals in the new database. In short, there were way too many people out there in the world with this variant for it to plausibly cause a rare disease like hypertrophic cardiomyopathy. We rechecked his heart and found that some of the thickness had subsided since he stopped competing. Athletic activity had been the explanation, after all! So we took away his diagnosis and released him; after two years, he was free, once again, to live his life to the fullest and even compete in track if he wanted. "I . . . I can run again?" he asked me, incredulous. "I can't believe it. I don't even know how to process this. I thought I would never have this opportunity again. I can have the love of my life back."

This is the power of population genetics for personalized, precision medicine—not that we needed to convince Theodore. He is now heading toward grad school to study . . . genetics.

All told, there are now more than sixty groups around the world that aim to enroll at least one hundred thousand people each in their biobanks. Meanwhile, work is under way led by the International HundredK+ Cohorts Consortium (IHCC) to connect each of these databases so that analyses can be carried out across them. There is also a Global Alliance for Genomics and Health (GA4GH, currently led by Ewan Birney, the creative, brilliant, and hilarious head of the European Bioinformatics Institute), devoted to enabling responsible genomic data sharing for the benefit of human health. With no more than a computer and an internet connection, researchers across the world will soon be able to access billions to trillions of data points, gaining insights from the health status and genomes of millions of global citizens. It is these insights that will help to demystify human diseases and identify targets for new drugs. Yet this is just the beginning of the realization of a dream that has captured the imagination of presidents and prime ministers, patients and participants, professors and the simply curious. In an era of personalized medicine, the most surprising thing, in the end, may be that it is understanding the population—All of Us—that is key.

19

# Genome Surgery

*We used to think the future was in the stars.*
*Now we know it's in our genes.*

—JAMES WATSON, NOBEL LAUREATE

*The power to control our species'*
*genetic future is awesome and terrifying.*

—JENNIFER DOUDNA, PH.D., *A Crack in Creation*

With a historical lens, it's not hard to understand why junior doctors are known as *residents*. As a junior doctor in the Southern General Hospital in Glasgow, Scotland, in the early 1960s, my father was on call every other day; he quite literally *lived* in the hospital. Thirty-five years later, when my turn came, junior doctor hours were beginning to be curtailed, but I would still not uncommonly be in the hospital from Saturday morning at 8:00 a.m. to Monday evening at 7:00 p.m. Today, residents' hours are more realistically compatible with a life, not to mention an apartment outside of the hospital, but these young men and women still routinely pull twenty-eight-hour shifts to provide continuity of care for acutely ill patients.

At Stanford, in the coronary care unit, we gather every morning to discuss the patients. The residents summarize how the prior day went for each patient, including their physical and mental state, and all the data from their various tests the day before. Everyone has his or her own style of presentation, and it takes only a few minutes to work out who has the magical touch: an ability to concisely summarize the key points and their implications, then formulate and present a coherent plan. The really good residents answer the questions in your head even before you ask them. But not everyone is born with an ability to pull the complex strands of a medical life story together and present a clear synthesis on minimal sleep. Some achieve this by neglecting other basic tenets of the hierarchy of needs. (One resident I remember came to work one morning wearing two differently colored shoes.) Occasionally, however, there is a master of the craft, a resident who goes above and beyond just presenting the story and making a plan.

I remember one such resident. It wasn't just the efficiency, clarity, and quiet confidence he displayed but the breadth and depth of his medical reasoning. His presentations were often laced with a soft ironic humor belied by an impish smile. One morning, in top-shelf parody mode—and in a triumph of insomnia-fueled, manic brilliance—he presented, in as best as I could tell one breath, a report on a patient who had suffered a heart attack that was delivered so fluently, so assuredly, so replete with small, relevant details backed at every turn by references to new and classic cardiology literature, that it took my breath away. This was next level. This was a doctor so effortlessly on top of his game that he had time to add flourishes and turn to smile at the camera as he passed the finish line. His game was not even cardiology. His name was Holbrook Kohrt.

I was not surprised to find out later that, in his chosen specialty of oncology (the care of patients with cancer), "Brook" was a star—a fast-track faculty recruit with an already flourishing research career. What I was surprised to learn later, however, was that he suffered from hemophilia, a rare genetic bleeding disorder. In the form of the disease from which he suffered, hemophilia A, the blood clotting factor VIII is missing, putting him at risk of uncontrolled bleeding into joints, muscles, the digestive tract, and most catastrophically, the brain. Clotting factors made by the liver cause the blood to clot by triggering cascading reactions, like dominoes falling. Lacking a critical component of the cascade, Brook had to avoid any kind of trauma. This was hard as an adult, but much harder as a young kid (he wore a helmet until he was seven years old to protect against blows to the head). When replacement artificial clotting factors were developed during his childhood years, Brook started to give himself injections of the missing factor VIII protein every two weeks. This can work for some patients for a while, but many patients relapse when their immune systems come to interpret the injected factor as foreign and start to make antibodies to neutralize it.

The close relationship Brook's treatment fostered with the medical system was part of what inspired him to go into medicine. "The connection I had with my hematologist was as close a connection as I had with my siblings," he once told *San Francisco Magazine.* And the brilliance that I witnessed over two weeks of treating heart patients together was evident in his research career, where he dedicated himself to researching ways to harness the immune system to fight cancer. There was irony in his choice of the immune system as savior, as Brook's own immune system eventually began to attack the foreign factor VIII he was injecting, limiting its effect and putting him once more at risk of serious bleeding. So this remarkable young scientist turned his formidable creative energy to his own fight for survival. He first identified which of his immune cells were making antibodies to the factor VIII, then worked to design a personalized therapy to disable them. As a result of this extraordinary journey, he was profiled in *The New York Times* and featured often in Stanford publications. Despite this attention, he held himself with a humility that endeared him to everyone he met. "Everyone's collaborator, everyone's adviser, mentor, and friend," his own mentor, Ron Levy, a colleague of mine at Stanford, once said of him. All of which made the news that, during a trip to Miami in February 2016, he died of complications of a bleed into his brain all the harder to bear. He was thirty-eight years old.

We understand, very well, the genetics of hemophilia. The form that affected Brook occurs mostly in males, because the gene for factor VIII is on the X chromosome. While females have two X chromosomes, meaning two chances for having a normal factor VIII gene to protect them, males only have one. Furthermore, advances in genome sequencing mean that we know for each patient exactly which mutation in their factor VIII gene causes the condition. So isn't it about time we were able to intervene directly at the site of the problem? Shouldn't our real ambition be to cure a genetic disease by correcting the genome's mistake back to normal? Although too late for this brilliant young doctor, incredibly, we are beginning to make progress on exactly that.

The idea of genetic therapy has been around since the early 1970s, but it wasn't until the early 1990s that the first human studies were performed. Unfortunately, these early attempts were largely ineffective and beset by problems. In the worst cases, toxicity related to the therapy led to highly publicized patient deaths, most notably that of eighteen-year-old Jesse Gelsinger in 1999 and to a group of children under treatment developing cancer. The popular press declared the end of gene therapy. "[Gelsinger's] death is the latest in a series of setbacks for a promising approach that has so far failed to deliver its first cure and that has been criticized as moving too quickly from the laboratory bench to the bedside," *The Washington Post* reported in 1999. Yet an NIH report on the failures was more practical. The expert panel concluded that we didn't yet know enough about the diseases targeted or the methods used to deliver genes into tissues, but that when we did, these could be effective interventions. Their message? That scientists should go back to the laboratory and learn more. The scientific community took note.

To be effective, a genetic therapy has to get to the cells that matter, get inside those cells, travel to the nucleus (where the genome lives), and in some cases, make changes to the genome itself. Delivering the therapy to the right place is the major challenge. In most cases, you can't simply inject DNA into a blood vessel and hope it goes where it needs to. Much of it would be destroyed in the bloodstream or in body tissues, and even if it did get to the right kinds of cells in your body, it would still be a long way from the nucleus, where it needs to go. Fortunately, nature has provided a very efficient method to deliver genetic material to cells, one you undoubtedly have personal experience with: virus infections.

No one really knows where viruses came from or how long they've been around, but these tiny protein balls containing DNA or RNA, one hundred times smaller than bacteria, have been recognized since the late 1800s. Given their primitive nature, they depend on cells to exist, forcing the cells they infect to copy them and make more viruses. Indeed, 8 percent of our own human genome is viral DNA that was left behind, so it is likely we coevolved with viruses. All of which makes viruses a useful vehicle for the delivery of therapeutic DNA. Early development of genetic therapies focused on designing viruses that could target and infect cells but not multiply in body tissues. The goal was to deliver a new, healthy gene directly to cells that needed it.

In the late 1990s, techniques were developed to purify enough virus for gene delivery to humans, something that required hundreds of liters of cultured cells. Not so much a laboratory as a factory. Later, in the early 2000s, more efficient techniques were ready for the first clinical trials, which were focused on patients with hemophilia. One reason hemophilia was chosen was that the cells that make the clotting factors are in the liver, and, it turns out, viruses home in on and infect the liver very well. The good news, especially given Jesse Gelsinger's death nearly a decade before, was that there were no safety concerns related to delivering a clotting factor gene using viruses. The bad news was that the effects were mild or short-lived, likely due to not delivering enough virus to the right tissue, or because the body's immune system ramped up a response to the virus in order to shut it down. Investigators were also surprised to learn that, although these were unfamiliar, artificial viruses, some patients already had antibodies capable of neutralizing them. The human immune system, it turns out, is a formidable foe. However, initial studies were promising, and now the race was on to make viral gene therapy better and more efficient.

It took ten years for the next major clinical trials in hemophilia to reach completion. These were aimed at hemophilia B—a condition in which the blood clotting factor IX is deficient (distinct from Brook's condition in which factor VIII was lacking). Researchers knew that it wouldn't take much additional clotting factor to make a difference. In studies where artificial factor IX was injected, major protective benefits appeared when clotting factor activity was just above 10 percent of normal levels. Initial attempts at gene therapy, where the virus was given with immune suppression to reduce the antibody response, produced an increase in factor IX activity from 0 to between 2 and 7 percent. This was promising, but short of the 10 percent where major benefits were expected. By the early 2010s, the discovery of a high-activity version of factor IX (known as *Padua*) meant that much lower doses of virus could be delivered with similar or greater effect. Indeed, studies in which factor IX Padua was delivered achieved clotting factor activity levels of around 34 percent. Not only was this three times the level needed for major therapeutic benefit, but the effect was sustained for months. This was a landmark for gene therapy, the first time a major and sustainable genetic correction had been achieved.

While sufferers of hemophilia B were understandably jubilant at this news, hemophilia A, the deficiency of factor VIII that Brook had, is six times more common. Yet it is also more challenging to treat via gene therapy. The factor VIII gene is twice as large as the factor IX gene. This matters because the viruses used for delivery, known as adeno-associated viruses (AAV), have a maximum load they can carry, and the factor VIII gene is too large. What to do? Could researchers use a different viral delivery system? How about shortening the gene in a way that still left its activity intact? Unlikely as it might seem, this latter approach actually worked. One company developed a size-optimized version of factor VIII, and, in a landmark study published in 2017, six of seven treated hemophilia A patients had *normal* clotting factor activity levels six months after gene therapy. These patients, who had been injecting synthetic factor VIII protein almost their whole lives, were able to simply stop. In the same period, their bleeding events dropped from an average of sixteen per year with the injections to one per year with the gene therapy. For them, this was truly miraculous. The lead investigator on this study, Professor John Pasi of the Haemophilia Centre at Barts Health NHS Trust in London, England, described the results in glowing terms uncharacteristic for a normally understated Brit. He called the results "mind-blowing," saying they "far exceeded our expectations." Nothing less than a road to a genetic cure had been paved.

Hemophilia is not the only genetic disease to have been successfully targeted this way in the last few years. As experience has grown, other readily accessible organs, such as the eye and the skeletal muscle, have also proven to be good targets for genetic therapy. The eye is particularly attractive because its interior is unusually well protected from immune system attack (something that could limit the effectiveness of viral delivery) and because diseases involving blindness are so devastatingly incapacitating.

Consider the Guardino family of Long Island, New York. Nino Guardino, an Italian master pizza maker, met his wife, Beth, a nurse, when she traveled to Italy on vacation. Together, they had a son, Christian, who, at six months of age, was diagnosed with Leber congenital amaurosis. This condition affects the retina, the light-sensing part at the back of the eye. The Guardinos were told that Christian was going to go completely blind. "It was devastating," his mom said. "I was so afraid of what it was going to do to him." As Christian's sight deteriorated, he began to navigate the world in dark glasses, holding a white cane.

The cause of Christian's blindness was a gene called *retinal pigment epithelium specific 65kDA protein* (*RPE65*). *RPE65* protein is essential for the eye to turn light into the electrical signals that it sends along the optic nerves to the brain. Without enough normal protein, those electrical signals never get sent, and vision becomes impossible. When he was twelve years old, however, Christian was given the chance to undergo a new experimental gene therapy from a company called Spark Therapeutics. The genetic therapy would add a working copy of the affected gene to the retinal cells at the back of the eye, allowing his body to replace the missing protein. The family signed up for the trial, and at age thirteen, Christian received a single injection to the back of each eye at the Children's Hospital of Philadelphia.

The results were dramatic. Over the following months, Christian experienced an 80 percent improvement in his vision. A few weeks after receiving the treatment, he turned to his mother and said, "Mom, is that you?" He had lost his vision so young that he had no memory of her face. "I remember seeing my mom and dad's face for the first time," he said. "Words can't describe that experience."

Throughout his ordeal, Christian had maintained his resilience through a love of music and, in particular, singing. Christian's story and his powerful singing voice were featured on the television show *America's Got Talent,* in 2017, where he wowed and inspired the judges and audience alike. Two years later, he told journalist Christopher Howard about what his newfound vision had brought him. "I have been able to see such incredible things since the gene therapy, like the moon, the stars, the sunsets, fireworks, snow falling—just so many things that I have had the opportunity to witness," he said. "And I can't take any of those things for granted."

Christian was not the only one in the trial to benefit. Indeed, in a group of patients with very limited vision who were headed toward complete blindness, the gene therapy improved their ability to navigate a maze and avoid obstacles in low light to such an extent that they performed tenfold better than the untreated control group, who deteriorated significantly over the course of the study. Fully 65 percent of the treated patients made the maximal possible improvement on the test.

In response to these dramatic results, the U.S. FDA approved the drug Luxturna (voretigene) in December of 2017. Scott Gottlieb, the FDA commissioner was enthusiastic: "I believe gene therapy will become a mainstay in treating, and maybe curing, many of our most devastating and intractable illnesses," he said.

Patients with retinal diseases are not the only ones to have realized the dramatic benefits of gene therapy in the modern era. Spinal muscular atrophy is a genetic condition that affects the motor nerves in the spinal cord. In the severest form of the disease, very young babies lose muscle tone (becoming limp or "floppy"), breathe poorly, and require ventilator support, which often results in pneumonia and death. It is one of the leading genetic causes of death in childhood. A less severe form presents later in infancy among children who are never able to stand or walk but who are able to sit up in a wheelchair. Although some of these children make it to adulthood, few live a normal life span. A still milder version of the condition is often diagnosed in the teenage years or even in adulthood, and many of those patients, despite motor challenges, can attain normal longevity.

Spinal muscular atrophy (SMA) is caused by mutations in a gene called *survival of motor neuron 1* (*SMN1*). The disease presents if both copies of the gene are affected (a recessive form of inheritance, meaning that very sick babies are born to unaffected parents). Until very recently, there was no effective therapy. The only available option was palliative care and mechanical ventilation.

This dire prognosis changed when a radical new gene therapy was approved in 2016. To understand how the new therapy works, we have to explore, for a moment, how a set of genes all contribute to producing SMN protein. In the human genome, there exists a closely related gene to *SMN1,* called *SMN2,* that can also produce SMN protein. In fact, some individuals have not just one but multiple copies of *SMN2.* Despite this, the human genome has come to rely entirely on *SMN1* for making SMN protein, suppressing protein production from *SMN2* by generating an "off" RNA message from the part of the genome that lies in between the two genes. Investigators found that when they inhibited this "off switch," the *SMN2* gene could be turned back on and co-opted into making functional SMN protein. This was a huge breakthrough. What if patients with broken *SMN1* could simply turn on their SMN2 to make the crucial SMN protein? Perhaps the genetic cure for spinal muscular atrophy was hiding in the genome in plain sight?

The approach to kill this "off switch" took the form of delivering a mirror-image version of the RNA message known as an *antisense oligonucleotide* directly into the fluid bathing the spinal cord (injecting directly into the spinal cord alleviates the need for a virus to deliver a payload). It was extremely effective. Out of 122 infants with the most devastating form of SMA who were given this treatment to turn on *SMN2,* more than 40 percent achieved normal motor milestones like rolling over, crawling, or even walking. This compared to none in the untreated group. There was also a corresponding decline in mortality. Indeed, by the end of the study, fewer than 40 percent of the treated infants had died, compared to almost 70 percent in the untreated group. This was simply unprecedented.

One downside to this particular form of therapy, however, is that repeat injections into the spinal cord are required every few months, because the antisense oligonucleotide is degraded by the body. In another approach, researchers have used viruses to deliver a fully functioning replacement gene to the brain and spinal cord by injection into a blood vessel in the arm. This approach, which is similar to the gene therapies previously described for hemophilia (that affected Holbrook Kohrt) and Leber congenital amaurosis (the disease that affected Christian Guardino), has shown even more promise. It was tested in children who presented with the severest form of the illness—a group who normally would never sit unaided, and most of whom would die before two years of age. After just one injection of the drug, nine out of twelve of these children were able to sit for more than thirty seconds, and two were able to crawl or even get up and walk. This was nothing short of miraculous. The results led to rapid FDA approval for Novartis for the gene therapy, Zolgensma (onasemnogene) in 2019. "Today's approval marks another milestone in the transformational power of gene and cell therapies to treat a wide range of diseases," the interim FDA commissioner Ned Sharpless said in a statement on the day the drug was approved. "The potential for gene therapy products to change the lives of those patients who may have faced a terminal condition, or worse, death, provides hope for the future."

It is clear from these groundbreaking examples that delivering a replacement gene can help correct a genetic disease when a gene is "broken" by a mutation and a necessary protein is absent. In some conditions, however, the problem is caused by *too much* of a mutant protein. In that case, you actually don't want *more* protein from a normal gene, you want *less* protein from an abnormal one. So what if you want to turn a harmful gene off? You could use the silencing approach discussed earlier to turn off an RNA message. But there is an even better way.

In 1998, Andrew Fire and Craig Mello, two researchers at the Carnegie Institution for Science's Department of Embryology in Baltimore, Maryland, reported that certain short, double-stranded RNA molecules (as short as twenty base pairs long) could shut down the expression of matched genes. They made this discovery in a 1 mm transparent worm called *Caenorhabditis elegans,* favored by geneticists. Although it was already known that similarly sized single-stranded RNA molecules could interfere with other RNA molecules with similar sequences (as described above for the *SMN2* story), this was inefficient and effects were short-lived. Moreover, the fact that so few double-stranded RNA molecules were required to trigger this effect in the worms suggested that a different mechanism was at play—one that could potentially be harnessed to shut down specific genes at will. Indeed, such was the excitement over the potential therapeutic applications in humans that Mello and Fire were awarded the Nobel Prize in Physiology or Medicine in 2006. What motivated the Nobel committee was not just the discovery of a fundamental new genetic process but the opening of a door to therapeutic gene silencing in humans.

The reason for the excitement was not hard to understand. As opposed to a strategy of delivering a whole new gene or using large amounts of an antisense oligonucleotide, as was the case with the then-current approaches, you might be able to treat some diseases by delivering much smaller amounts of short, double-stranded RNA molecules. Predictably, the excitement led to hype, and not unlike a decade prior when gene therapy was first attempted, several companies were formed in the early 2000s to exploit the potential of RNA silencing (also known as *RNA interference,* or *RNAi*) to treat human disease. Once again, however, early hopes were dashed as challenges in delivering the therapeutic RNA molecules to the right cells in the body plagued the RNA-silencing drug development programs. By the early 2010s, investors and large pharmaceutical companies were running scared. "People started giving up hope," John Maraganore, chief executive of Alnylam Pharmaceuticals remembered. Scientists, however, kept working toward a better understanding of how this fundamental process worked and, critically, how best to harness it safely and target it precisely to the organs of interest. In particular, delivering the RNA by packaging it into tiny, protective nanoparticles began to show promise, especially for targeting cells in the liver.

In August 2018, the first RNA-silencing drug packaged into nanoparticles was approved by the FDA. The target, in this case, was hereditary transthyretin amyloidosis, a devastating disease that causes problems with nerve sensation, bowel and bladder function, heart rhythm, and heart muscle. It was also a disease without any effective therapy. Most patients presenting with heart symptoms die of heart failure within two to three years. The cause is a mutated version of the *TTR* gene. This creates a mutated transthyretin protein, and it also causes the normal transthyretin protein to be misfolded in ways that lead to toxic accumulation in tissues such as those of the nervous system and the heart. The RNA-silencing drug—Onpattro (patisiran), made by Alnylam Pharmaceuticals—was designed to reduce levels of both mutant and normal transthyretin protein. In 2018, a clinical trial in 228 patients showed an unprecedented impact on disease progression. Indeed, over eighteen months, as the disease progressed in patients who received the placebo and their walking speed declined, the patients in the treatment group got faster. This was a drug that appeared not just to arrest the progression of the disease but actually to reverse its course.

Furthermore, the promise of gene-silencing therapy is not limited to rare genetic disease. Consider, for example, its application to stubborn high cholesterol. In August 2019, results from a landmark study were presented at the European Society of Cardiology meeting in Paris. The study focused on patients whose LDL levels remained high despite maximal doses of cholesterol-lowering medication. Among the 1,617 participants from seven countries, the average starting LDL was 107 mg/dL. (Remember, the target for patients who have had heart attacks is 70 mg/dl or below.) While the patients who received a placebo showed an overall 4 percent increase in their LDL over the eighteen-month trial, the LDL in patients treated with the RNA-silencing therapy dropped by a startling 49 percent. In chapter 17, I introduced Sharlayne Tracy and her superhuman low cholesterol levels—the result of being born with two inactive *PCSK9* genes. Drug companies successfully mimicked her natural superpower by using antibodies against *PCSK9* to inactivate the protein. However, you might recall that this cholesterol-lowering approach required new injections of antibodies every two to four weeks. The RNA-silencing therapy, by contrast, needs to be injected just twice per year.

My own interest in RNA silencing began in 2006. I had recently joined the faculty at Stanford and was interested in how we could do better than to just describe the genetic diseases that cause heart failure and sudden death. Surely we could use that knowledge to actually treat the disease? While I understood that completely silencing a gene, as described above for cholesterol management, could be effective for some conditions, that clearly wasn't going to be the right approach in inherited cardiomyopathy, where only one copy of the gene is affected and the other copy is critical to the functioning of the heart. Could we find a way to knock down just the one mutated copy? That would be a really elegant approach, it seemed to me. I had heard about RNA silencing, and that seemed very promising. There was just one problem: I didn't know anything about RNA silencing.

Fortunately, there was someone at Stanford who knew quite a lot: Andrew Fire, the Nobel Prize winner. He had moved to Stanford from Johns Hopkins University in 2003. Hearing that he was approachable, I spent some time carefully crafting what I hoped was an intriguing email: "Dear Professor Fire: You don't know me. I'm a new junior faculty member here," it began. Then I proceeded to politely ask if there was any chance he could spare a moment to give me some advice—any time, well, in the next few months. My calendar was wide open, I noted, while I tried to communicate that I would go anywhere, anytime, for a few moments of his time. I sent the email off, hopeful, but also understanding that famous Nobel Prize–winning professors are busy people who receive requests and invitations on an almost daily basis. I had moved on to other things when, a few minutes later, my phone rang. I should explain that in the office I had been given as a new faculty member, there was a landline phone and, somehow, I had managed to inadvertently adopt the phone number of an emeritus faculty member who previously ran the cardiac rehabilitation program at Stanford. This meant that more than 90 percent of my incoming calls were wrong numbers, and I got used to patiently listening for the right moment to interrupt and recite the actual number of the person they were looking for. So it was in that light that I picked up the phone that afternoon, already glancing at my yellow sticky note bearing the number for the emeritus professor.

"Hello," I said.

"Hi, it's Andy," the voice on the other end of the phone replied.

I stopped in my tracks. *Andy?* I thought. *I really don't know any Andys.* I didn't recognize the voice on the phone, either. "Sorry," I said, "the number I think you're looking for is—"

The voice interrupted me. "Andy Fire. I just got your note. Do you want to come over?"

I was dumbstruck. After I clarified that he did actually mean "now," I found myself, barely hours after having typed out the idea for the first time, in the office of the man who won the Nobel Prize for discovering the process I was hoping to co-opt. It was inspiring to me that such a renowned scientist was willing to take the time to talk to a junior colleague. And it led to, so far, twelve years of experimentation and collaboration aimed at harnessing RNA silencing as a therapeutic approach for cardiomyopathy.

We took a stepwise approach. We first had to convince ourselves that we could silence "just" the mutated copy of the gene. The two genes we chose to focus on were *MYH7* and *MYL2,* both of which caused hypertrophic cardiomyopathy. So the first step was to put both a normal and a mutated copy of the gene in a cell type that we could easily keep alive in a dish (since heart cells are tricky, we often start with kidney cells). Matt Wheeler and postdoctoral scholar Kathia Zaleta attached a "green light" to the normal version of the heart gene and a "red light" to the mutant version, then proceeded to test different RNA-silencing fragments to see if we could selectively quiet just the mutant. We detected an effect by looking for changes in the amount of red and green light coming from the cells. Some of these "therapies" worked better than others, and one or two worked very well indeed. We took the one that worked the best and tried to see if it would also work in heart cells from rats with cardiomyopathy. It did! Then we were ready to try our new therapy in a live animal. A very talented microsurgeon in my lab injected our RNA-silencing treatment, now packed into a virus, into the tiny veins of newborn mice destined to develop cardiomyopathy by virtue of having both mutant and normal copies of the gene *MYL2*. These baby mice are very small—about 1 cm long and weighing about 1 g—so injections are performed with great care. We then watched these little babies carefully as they grew into teenagers. We had added a special light-emitting gene inside the virus payload so we could check that our viruses really made it to the heart and not anywhere else. We then followed our furry patients until they were adults, and we carried out ultrasounds of their hearts (which beat somewhere between six hundred and eight hundred times per minute). We also ran them on mini treadmills, just like our human patients. Those who had received the RNA-silencing treatment showed clear evidence of a reversal of the disease! Their hearts were less thick, less stiff, and when we looked in detail at the tissue under the microscope, their hearts looked much more like normal hearts than those from the untreated mice. Most importantly, the treated mice survived longer.

But that was mice. They have teensy hearts that beat up to one thousand times per minute when they're excited. The walls of their hearts are just 1 mm thick. Their biology is not the same as humans. We needed to think about whether our treatment would work in humans as well.

In developing a drug, you can't go straight from mice to humans, of course. So as a start, we wanted to try our RNA-silencing therapy in human heart cells in a lab dish. The only problem is, heart cells are not readily available. Removing a piece of a patient's heart is rarely performed, and cells from hearts that are removed during transplant don't survive well, because the heart that is removed is so sick. In addition, what we want are cells that genetically match the disease or patient we are focused on. We therefore took advantage of another Nobel Prize–winning breakthrough: the ability to turn the clock back on cells to make them stem cells again. Stem cells are abundant in the early embryo and remain available in some parts of the body throughout our entire lives. What makes them unique is their ability to turn into any kind of cell. Turning the clock back means we can take a blood sample from a patient, turn those blood cells into stem cells, then persuade those stem cells to turn into heart cells—heart cells that have exactly the genetic change we want to study. This magic is achieved by using a cocktail of chemicals, and the result is a formation of beating cells—a kind of mini heart—in a dish (though it is not shaped much like a heart, more like a blob). Alex Dainis, a graduate student in my lab, was focusing on dishes of these mini hearts as well as single isolated cells grown from the blood cells of two sisters—patients of ours with a severe form of hypertrophic cardiomyopathy (in fact, the same form as the family with the Coaticook curse described in chapter 15). We wanted to know: Could the RNA-silencing therapy work in real human heart cells genetically predisposed to develop disease? Would it shut down just the mutated gene and not the much-needed healthy one? If so, that would be a key step toward our goal of treating real patients. Alex measured the size, shape, contraction, and relaxation of the heart cells (patients with hypertrophic cardiomyopathy have hearts that are thick, are stiff, and contract too vigorously, and the cells in the dish reflected those changes). She made her measurements before and after delivering the gene-silencing therapy. It worked! She found that her therapies could selectively silence just the mutant gene, change the molecular markers in a positive direction, and reduce the force of contraction of the cells back toward normal.

Thanks to the hard work of scientists like Alex all around the world, new drugs based on the RNA-silencing approach continue to emerge.

It is often said that the top three challenges in gene therapy are delivery, delivery, and delivery. The point is, it is usually much harder to get the therapy where it needs to go than it is to design the therapy itself. Ready accessibility is one reason that blood cells and blood diseases have been attractive targets for gene therapy.

You may have heard of "bubble boy" disease, otherwise known as severe combined immunodeficiency (SCID). This X-linked disease affects boys, who have just one X chromosome, and causes abnormal development of both arms of the immune system (hence the term *combined*—it affects T cells and B cells). It first captured the public imagination because of David Vetter, a boy born in 1971 with SCID who spent most of his life in a plastic sterile "bubble" avoiding infection, waiting for a bone marrow transplant. Such a transplant could be curative because a donor's transplanted cells with their different genomes would have a functioning copy of the causative gene (usually the gene *IL2RG*), allowing a fully functional immune system to form.

In a typical bone marrow transplant, usually performed as a treatment for myeloma or leukemia, the patient's own immune system and blood-forming cells are first shut down using chemotherapy or radiotherapy. Then new blood-forming stem cells are infused either from the patient or from a donor. It is a harrowing and extremely risky procedure, which throws the patient into a state much like David Vetter's. Shutting down the immune system, even temporarily, puts the patient at risk of serious infections. A subsequent challenge is *graft versus host disease,* which is, as the name would imply, when the new immune system cells from the donor graft start attacking the host tissues, treating them as foreign invaders.

Given these challenges in receiving bone marrow from another person, an attractive approach for genetic diseases of the blood and immune system would be to remove the patient's own bone marrow stem cells, genetically fix them, then return them to the patient. It still requires killing off the old cells, but now instead of someone else's cells, normal cells can be infused that are perfectly matched to the patient because they are the patient's own cells. The idea has tantalized researchers for years.

In the early 2000s, however, attempts to treat SCID with this type of gene therapy were beset with fatal problems. Because the gene therapy unintentionally disrupted other genes, the therapy itself led to cases of leukemia, a low point in the stop-start history of progress in genetic therapy. A decade of work ensued to understand what caused the leukemia, and a study published in April 2019 illustrates how far we've come. After removing bone marrow from eight infants born with SCID who had low numbers of functioning immune cells, chronic infections, and slowed growth, the researchers used a specially adapted virus to add a new copy of the defective *IL2RG* gene to each infant's bone marrow stem cells before returning the treated cells to the babies. In addition, they gave a chemotherapy drug to reduce the number of immune cells being built in the babies' own bone marrow. Remarkably, after three to four months, the numbers of functioning immune cells were in the normal range in seven out of the eight treated babies, while the eighth achieved normal counts after just one more boost of treated cells. The boys' lingering infections cleared very quickly, and they began to grow normally. This was a revolutionary cure that is now working its way toward FDA approval.

The success has also led to interest in treating more common blood diseases, including genetic conditions involving hemoglobin, the molecule that transports oxygen inside red blood cells. The best known of these hemoglobinopathies is sickle cell disease, a condition that affects mostly those with African or Hispanic ancestry. Three hundred thousand children are born each year with the disease, which causes crises of excruciating pain when red blood cells warp into a crescent (sickle) shape and get caught on each other and stuck inside blood vessels—impeding the flow of oxygen to tissues. Carmen Duncan, a twenty-year-old from South Carolina, was one of those patients. She spent her childhood in and out of hospitals, racking up tens of thousands of dollars of health care costs. She even had to have her spleen removed when she was two years old because of damage caused by the abnormal red cells clogging up blood flow in that highly vascular organ. Her crises could last weeks, and she would ache all over. "A simple touch really hurt," she told Gina Kolata of *The New York Times.* But after receiving a onetime gene therapy called Zynteglo (LentiGlobin) from Bluebird Bio, she no longer has any signs of the disease at all. The therapy delivers a working copy of the hemoglobin gene that is modified in a way that helps stop the sickling effect of the mutant hemoglobin to the patient's own stem cells. A picture of Carmen, pain-free and smiling, appeared in the January 27, 2020, edition of *The New York Times.* Meanwhile, the Bluebird Bio therapy, along with drugs from other companies aimed at sickle cell disease, is heading toward FDA approval.

Genetic therapy has achieved remarkable success in the last few years by delivering an extra copy of a normal gene or by eliminating a troublesome gene's effects through RNA silencing. But what if it were possible to surgically correct the genome, to edit a mutated sequence back to normal? Such genome surgery represents the holy grail of genetic therapy.

Techniques to someday enable this kind of gene correction began development in the 1990s, and progress accelerated in the early 2000s. Initial work was based on the idea of making a break in the DNA near the mutation, then allowing the cell to repair the break using the non-mutated copy of the gene. Yet breakthroughs in the decade starting in 2010 changed this landscape completely. In particular, the discovery of a bacterial defense system called CRISPR (pronounced "crisper" and short for *clustered regularly interspaced short palindromic repeats*) enabled researchers for the first time to target specific areas of the genome and repair them. It catapulted gene editing to the front pages of newspapers as the hype machine rolled in, accompanied by bicoastal patent wars between Harvard University and the University of California at Berkeley.

The CRISPR story has its origins in the 1990s with the finding of unusual repeated sequences in bacteria by Francisco Mojica, a graduate student at the University of Alicante. He spent years studying these unusual sequences in a wide range of bacteria. The sequences he found were typically about thirty letters long, repeated, often palindromic (a palindrome is a word spelled the same backward as forward, like the word *rotator*) and separated by DNA "spacers" of about thirty-six base pairs. He was unclear at first what the purpose of the system was, but after finding those spacer sequences matched the sequences of certain viruses, and in some cases, viruses that the bacterial strain was resistant to, he deduced that CRISPR sequences were part of a bacterial defense system against viruses. They were, in effect, a form of bacterial "memory."

Some years later, in the years leading up to 2010, a Lithuanian scientist, Virginijus Šikšnys, at the Institute of Applied Enzymology in Vilnius, Lithuania, demonstrated that it was possible to transfer a CRISPR system from one bacterial strain to another and confer viral resistance. Around the same time, in 2011, a meeting took place between two RNA scientists at a symposium in Puerto Rico. Emmanuelle Charpentier, a Viennese scientist at the Laboratory for Molecular Infection Medicine in Sweden, met Jennifer Doudna, a Hawaiian RNA biologist from the University of California at Berkeley. Together, Doudna and

Charpentier characterized the "molecular scalpel," a molecule called *Cas9* that could cut DNA (cutting DNA was an important part of the defense mechanism against viruses). Because human cells have "on-demand" repair molecules that swoop in and repair cut DNA, Doudna and Charpentier reasoned that if you could direct the molecular scalpel to the spot where a mutation existed, and then provide a "normal" piece of DNA as a repair template, you could provoke the cell into repairing its own genome. It was brilliant, groundbreaking, new, and yet built on a biological mechanism millions of years old. In 2012, two teams, one led by Šikšnys and another by Charpentier and Doudna, published papers demonstrating the potential to manipulate the gene-editing system in bacteria to edit DNA.

Meanwhile, two scientists from Harvard homed in on the potential of CRISPR for gene therapy in humans. George Church, contrarian, polymath, and world-renowned geneticist, and Feng Zhang, a rising star assistant professor at the Broad Institute, worked both independently and together to show in separate papers that gene editing of human cells using the CRISPR-Cas9 system was possible and powerful.

The foundational work of Mojica and Šikšnys in describing the system, the creative collaboration of Charpentier and Doudna (recognized with a Nobel Prize in 2020) in characterizing Cas9, and the work of Zhang and Church to apply it to human cells laid the foundation for a genome editing revolution. No longer would gene therapy be limited to delivering new copies of genes or silencing the effects of mutant genes. Now there was the potential for editing the genome back to normal.

The science around CRISPR continues to advance toward greater precision and higher efficiency. David Liu's laboratory at Harvard has used the Cas9 system to target a specific area of the genome for direct genome editing, without cutting. Instead of "cut and repair," his base editors swap one letter for another, much like a word processor would allow you to delete one letter and replace it with another (rather than chopping up the sentence and swapping in a whole new sentence minus the spelling error, as was the case with the original Cas9 approach). The newest version of his technology can go beyond editing a single base to precisely inserting almost any replacement piece of DNA, again without a cut. Liu has calculated that by using this technique, almost 90 percent of genetic variants associated with human diseases could be corrected.

With the science of gene editing advancing, it was inevitable that scientists would aspire to correct disease mutations not just in human adults or in cells from their bone marrow but also in human embryos. One scientist made headlines in 2017 when his team corrected a mutation causative of hypertrophic cardiomyopathy in human embryos. Shoukhrat Mitalipov, from the Oregon Health & Science University in Portland, created human embryos by fertilizing donated eggs with the sperm of a man suffering from hypertrophic cardiomyopathy (the condition that affected Leilani Graham, whom we met in chapters 15 and 16, and that we focused on earlier in this chapter using RNA silencing). Since only one copy of the *MYBPC3* gene is affected in this disease, and each sperm has a fifty-fifty chance of carrying it, it would be expected that—if there was no effect of the gene editing—half of the embryos would suffer from the disease and half would not. In Mitalipov's study, published in the journal *Nature* in August 2017, he showed that forty-two of the fifty-eight embryos had two normal copies of the gene: in other words, far more than half. His paper was followed by some controversy. For example, some scientists were not convinced that the CRISPR editing worked as billed. Although Mitalipov provided a normal copy of the gene sequence as a template for the cell to use, data in the paper suggested that the embryo cells might have instead used the other, healthy copy in their own genomes to repair the CRISPR cut. This provoked skepticism in the community. Scientists also wondered if, instead, the CRISPR system had simply cut out a chunk of the mutated gene, making it appear as though it had been repaired (because the mutant was now missing) whereas, in fact, only one normal copy was present. In response, the following year Mitalipov's team provided new data reinforcing their assertion that the CRISPR editing had worked as described and answering some of these questions, but to date, both the approach and the mechanism remain controversial.

The most heated debate, however, focused on the act of intervening on human embryos itself—specifically, the ethics of making permanent genetic changes to the human species, changes that can be passed from future generation to generation. Whereas gene editing in a specific tissue or organ of an adult patient affects only them, editing an entire human embryo means that those changes will end up in the individual's sperm or eggs and be passed on to future generations. Mitalipov was extremely clear that the embryos he studied were never going to be implanted into a mother to grow into humans. One researcher in China, however, He Jiankui, did not observe those boundaries. With an eye toward a page in the history books, he announced in November 2018 the birth of the world's first gene-edited babies. His motivation was to edit a gene called *CCR5* to, he claimed, protect the children from HIV infection. It seems, however, that the page in the history books he has most certainly garnered for himself does not feature the glowing words of praise he had imagined. Jiankui was immediately condemned for his naïveté and labeled as dangerous. A worldwide press outcry sent him into hiding, and in late 2019, it was announced that a secret criminal trial in China had led to a sentence of three years in prison. Questions about what the parents of the gene-edited infants had been told, about the efficacy of this mechanism of protection from HIV, and questions relating to the safety of the gene-editing technique itself culminated in a tsunami of almost universal disapproval. Jiankui's work spurred professional societies to work together to publish a position paper calling for a moratorium on the use of gene editing in human embryos. Francis Collins, the head of the NIH, called it "deeply disturbing" and "profoundly unfortunate," suggesting that should such "epic scientific misadventures" proceed then "a technology with enormous promise for prevention and treatment of disease will be overshadowed by justifiable public outrage, fear, and disgust." Strong words, indeed. With the memory of the failures of genetic therapy from the 1990s still fresh, there was universal agreement that more science is required to demonstrate the safety of gene editing before it would be ethical to consider editing the genomes of embryos destined to grow into children.

We have discussed in this book the transformative potential of reading the genome to diagnose disease. In this chapter, I have described the exciting new technologies that are allowing us to act on that information. What could we achieve if we brought those technologies together?

Mila Makovec loved the outdoors. She was an active and communicative toddler, but as she grew, her mom started to notice she was talking less. Also, her movements seemed less coordinated, and her vision was deteriorating. By age five, she was blind and unable to stand. At times, she could not support her own head, and she suffered dozens of short seizures every day. Like many undiagnosed patients, the family traveled from one doctor to another trying to find answers. As her symptoms progressed, the diagnosis came into focus: Mila was suffering from Batten disease, a degenerative neurological disease known to be caused by mutations in both copies of the gene *CLN7*. Much worse, Batten disease was universally fatal.

Mila's mom, Julia, started researching the condition and was puzzled by the fact that her geneticist had found only one variant in *CLN7* in Mila, one that came from Mila's father. Normally, in Batten disease, both copies of the gene need to be affected. Hoping someone could help to paint a fuller picture of the genetics of Mila's disease, she reached out through Facebook in January 2017, asking if anyone in her network could suggest someone who might be able to help with whole genome sequencing. That request was seen by the wife of Timothy Yu, a pediatric geneticist and neurologist at Boston Children's Hospital.

Less than a month later, Yu found himself looking at the whole genome results for the family. He found the variant that Mila had inherited from her father quite quickly, but that variant alone really couldn't explain her presentation. He looked deeper—including, since he had the whole genome, at the space in between the genes. There, Yu noticed a piece of DNA that shouldn't have been there. It was two thousand letters long and a "jumping gene" (a sequence of DNA that can move from one position in the genome to another). It seemed clear this jumping gene was somehow affecting the expression of the maternal copy of Mila's *CLN7* gene. The mystery appeared solved.

But Tim Yu wasn't done. He was familiar with the RNA-silencing drug for spinal muscular atrophy described earlier. He wondered if this same approach could work in Mila's case. Could he turn off the signal from the jumping gene and allow Mila's otherwise normal gene to do its job?

Certainly, it seemed theoretically possible, but designing a drug was not something he had done before. In fact, designing a drug for one single patient was not something anyone had done before. He started talking to anyone who could possibly help—geneticists, pharmacists, drug developers, regulatory experts, hospital managers, ethicists, and more. Remarkably, in less than six months, he and his collaborators had developed a prototype drug that they were ready to try—one designed to shut down the effect of the jumping gene. First, they applied the treatment to Mila's cells in a dish. Excitingly, the effect of the jumping gene seemed to have been silenced; they found normal levels of the *CLN7* RNA message. But before they could proceed in giving the treatment to Mila herself, they needed approval from the FDA.

Over the winter, Mila's clinical state worsened. Her seizures increased, and her motor functioning was deteriorating. It was now or never. In January 2018, under a special pathway called *compassionate use,* Yu's team got approval from the FDA for their drug, which they named Milasen and injected it into the fluid around her spinal cord. It was just one year since her Mom's Facebook post.

Before the treatment, Mila was having up to thirty seizures per day, each one lasting a couple of minutes. After the therapy, the seizures became shorter and rarer. Her parents also noticed she was communicating better. In a disease as universally progressive as Batten disease, spontaneous improvement is unheard of. The only possible conclusion was that the therapy was doing its job.

From genome sequencing to a personalized genetic therapy in just over a year: Yu's achievement was a tour de force of dogged determination and seat-of-the-pants drug development carried out at breakneck speed. It isn't an approach that will work for everyone, and it was certainly expensive (funded mostly by family fundraising). But as an example of the power of truly personalized therapy, it stands unparalleled.

We are in a golden age of genetic therapy. After initial setbacks, there are now hundreds of gene therapy programs at centers throughout the world. Every few months, we witness major new breakthroughs both in the basic science of manipulating the genome and in clinical trials testing these breakthroughs in patients. The speed with which we can move from diagnosis to therapy has changed dramatically. After decades of concentrated effort, we are finally learning how to harness natural processes honed over millions of years of evolution to translate our unparalleled ability to read the genome into a new opportunity to write it and to correct the grave mistakes that turn it against us.

# 20

# The Road Ahead

*Most people overestimate what they can do in one year*
*and underestimate what they can do in ten.*

—WILLIAM HENRY GATES III,
COCHAIR OF THE BILL AND MELINDA GATES FOUNDATION

*You can't do two, until you've done one.*

—ANONYMOUS

It is more than a decade now since I first went to see Steve Quake in his office and watched agog as he scrolled through his own genome on the screen. In those intervening years, much has changed. The global community's collective exploration of the human genome, and its impact on medicine, has not slowed down. In fact, the pace appears, if anything, to be quickening. The next few years should see us able to access the full contents of millions of human genomes using little more than an internet connection. And every day, children and adults will continue to have their diagnostic odysseys ended thanks to newfound powers to decipher the genome.

Yet it didn't happen instantly or by chance. When we started work on the Quake and West family genomes, back in 2009, it seemed like any day we would be sequencing everyone—unlocking medical secrets that would change all our lives. Yet even in times of apparently revolutionary change, that change takes time to fully come to fruition.

There certainly were barriers to the advancement of genomic medicine: healthy skepticism about new technology, government regulation, privacy concerns, and hesitation from the health systems being asked to foot the bill. However, steadily, as our community addressed each of these issues, the future crept closer—until we turned around one day and realized that the future had already arrived. The genome has gone from a billion-dollar, multiyear, multicountry proposition to an everyday part of the practice of medicine. Today, a physician can order a genome for a rare-disease patient almost as easily as ordering a cholesterol test. Health insurance companies increasingly list it as a covered benefit, acknowledging that transformative insights can emerge. Some health systems are even starting to offer genetic sequencing as part of preventive care—a way to reveal disease risks in advance of the disease arising.

So what now does the future hold? For starters, sequencing technology will continue to improve. Yes, genomes will get even cheaper and faster to produce. But much more important, the genome data will also get better. Accuracy will improve, and we will start to shine a much more powerful light into the dark corners of the genome. In chapter 14, long-read sequencing technology from a company called Pacific Biosciences helped us find a disease-causing deletion in Ricky's genome that traditional methods had missed. That technology can read stretches of DNA that average ten thousand letters long (as compared to a few hundred letters for Illumina's technology). As these longer "reads" become more common, assembling a genome will grow ever easier. (Remember, a jigsaw with only ten pieces is easier to put together than a jigsaw with one thousand pieces.) Important structural changes, like DNA deletions and duplications within a gene, will be spotted more easily. And in analyzing inherited traits within a family, it will be ever easier to work out which pieces of DNA come from Mom and which come from Dad.

Pacific Biosciences' technology isn't the only approach to getting long DNA reads. Another exciting technology produces long DNA reads using a special protein called a *nanopore.* If you line up lots of these nanopores and add DNA or RNA, those molecules will start to fall through these pores. It's as if each one is a long thread being pulled through the eye of a needle. As they pass through the "eye," a tiny electrical current is measured across the pore, and that current changes depending on which letter of DNA or RNA is passing through at that moment. As the electrical signal fluctuates, the sequencer generates a series of squiggly lines that look like a badly scrawled signature. This output is even called a *squigglegram.* And computer programs can translate these squigglegrams into a real-time readout of DNA or RNA letters as they pass through the pore. These nanopore reads can be really long. In fact, sometimes a DNA molecule as long as two million letters transits through the pore! With enough pores, you could read a genome in a few hours (contrast almost a day for the Illumina technology). Eventually, this will be cut to minutes.

The current leader in nanopore sequencing is a company called Oxford Nanopore from Oxford, UK. Its chief technology officer is none other than Clive Brown, whom we met in chapter 4. (He was the English bioinformatician who joined the Cambridge, UK–based sequencing company Solexa, which was later sold to Illumina.) Clive and the nanopore team are now the grand disruptors of a sequencing industry that has been dominated by Illumina for the past fifteen years. They have introduced products like a portable sequencer less than half the size of a typical smartphone. In contrast with the hundreds of thousands of dollars required to acquire an Illumina or Pacific Biosciences sequencer, they have essentially given away these miniature sequencers for free, charging scientists for the chemicals required to run them rather than for the machines themselves. Many groups that could never otherwise afford a sequencer can now do sequencing at their lab bench, in a remote jungle, or even in outer space. (Stanford grad and astronaut Kate Rubins recently used a nanopore sequencer to sequence virus, bacteria, and mouse DNA while aboard the International Space Station.)

Being able to sequence anywhere opens up new possibilities for rapid diagnosis in the field. Consider, for example, the protection of food crops from disease. Plant viruses cause tens of billions of dollars in damage annually, leaving millions of people at risk of food insecurity. The cassava plant, a carbohydrate crop that is a major source of calories for over eight hundred million people worldwide, is grown by resource-poor farmers in sub-Saharan Africa. This small, portable sequencer allowed scientists to identify—in under four hours—the cause of the widespread destruction of their cassava plants as the begomovirus. This knowledge allows farmers to quickly take action to protect their livelihoods through controlling the whitefly that spreads the virus, or planting virus-free or -resistant cultivars. In the United States, nanopore technology has been used to detect, in just a few hours, salmonella in shrimp from India and *E. coli* in beef from the United States. Clearly, putting sequencing technology to work local to where the outbreak arises can dramatically cut the time to identify the source of infection and limit its harmful effects on human health.

Long-read approaches also provide more information than "just" the letters of the DNA code. DNA, like all molecules in the body, can be altered by chemical changes that control which genes are turned on or turned off. Understanding these chemical changes is part of a field called *epigenetics*. Typically, detecting these DNA modifications requires a special kind of gene chip or an alternative method of sequencing, independent from the standard sequencing approach. The long-read approaches we have discussed, however, are able to pick up many of these chemical changes at the same time as reading the base letters of the genome (by detecting different patterns in the squigglegram, for example), saving both time and money. This additional layer of information will, over time, help us identify which gene mutations are most important and will likely reveal new kinds of disease-causing genome changes that can either shut down or overamplify the expression of a gene.

So in the future, technology will bring us more and better genome information. How will this help us understand disease?

The diseases we have talked about in this book are mostly rare. (Collectively, of course, rare diseases are rather common—one in fifteen people has a rare disease.) The question I get asked most often by the public is: "When is the genome going to be ready for me?" After all, even the fourteen out of fifteen people who evade a rare disease will one day age, take medications, and most likely suffer from a common disease such as heart disease or cancer. When will the genome be something your general practitioner consults as part of your everyday care?

For Steve Quake, the West family, and other early adopters in Stanford's primary care clinic, we did our best to provide predictive scores for a range of common diseases and advice on personalized medication prescribing. Those predictive scores were not yet ready for prime time. They were based on studies of mere thousands, rather than hundreds of thousands, or even millions, of people—and were not quite accurate enough. Yet it was clear that with every year that went by, those scores would improve until at some point they would be accurate and relevant enough for everyone. Excitingly, that day has finally come. We have known for years that about half of the risk for heart attack is nature (your genes) and half is nurture (your behavior or environment). Yet some standard risk scores used around the world today don't include any questions about family history that might speak to the nature part. When we add genetic risk scores into the mix, not surprisingly, we find that our predictions are more accurate. That means some people move up and some move down the risk table "league standings." It also means that when we don't incorporate genetic risk scores, we are missing the chance to help some people who need it and we are overtreating others who don't. So what about the cost? In fact, to get enough genetic information to calculate these scores, you do not need to sequence each spot in the genome thirty or forty times like you do to diagnose rare disease. You can cover each spot just once, making it many times cheaper than fully sequencing the genome. We're talking now about a test that can estimate risk for heart disease, cancer, and dozens of other diseases for less than most people pay for a haircut.

One important consideration, as we begin to incorporate genetic risk scores into medical practice, however, is the disparity in how well these estimates perform across those with different ancestry. Genetic risk scores are mostly generated from research involving people of northern European ancestry. Not surprisingly, they perform less well in people whose ancestry derives from other parts of the world. Should we wait to deploy them, then, until we have sequenced hundreds of thousands more people from every ancestry group? Absolutely not. We should, of course, urgently do those studies in diverse populations to allow us to calculate the best scores for everyone based on their unique ancestry, but since improved prediction is available for everyone now, and those underrepresented in studies are often the most in need, we should start to offer our most accurate predictions to everyone right away.

Better prediction is, of course, only useful if we have interventions that can reduce the impact of the disease. Is that true for heart attack? It is probably not surprising to hear from a cardiologist that everyone should eat better and exercise more. But equally important is that such lifestyle changes can counteract the risk coming from both nature and nurture. In addition, we have medications for high cholesterol and high blood pressure that we know can save lives. The question is: Who is at greatest risk and who can benefit most from these interventions? At Stanford, in 2021, we started to include genetic risk information, based on studies with hundreds of thousands of patients in cardiovascular clinics, to make our predictions more accurate. We are hopeful to be able to report better outcomes for our patients in the next few years.

And it isn't just heart disease. We are better at incorporating a family history of cancer into our screening recommendations for cancer than we are for heart disease. We also know, however, that because of gaps in knowledge and selective memory, asking someone about their family history is a poor guide to their actual genetic risk. In breast cancer screening, an estimated increase in risk over the average for an individual of greater than 25 percent leads to a recommendation to use magnetic resonance imaging (MRI) of the breast rather than only x-ray mammography. The genome can help guide us as to who should receive this more intensive screening.

The future also lies in real-time monitoring of our health. Consider that jet engines, responsible for safely transporting millions of humans each year at thirty thousand feet, send terabytes of data *per hour* back to airline engineers so that problems in the motors can be detected early. Cars are now equipped with sensors that predict a collision and slam on the brakes before it occurs. Yet in medicine, when our focus is the most important "engine" of all, we make do with an annual checkup (which many people skip) and a one-size-fits-all approach to screening. In fact, the medical profession considers *screening* a bit of a dirty word, because historically, we have been so bad at accurately interpreting test results, each additional test risks flooding the health care system with an unmanageable number of false positives. So we wait until someone develops a hacking cough before doing the chest x-ray that would diagnose their lung cancer. Or we wait until a blood clot travels to the brain, causing a stroke, before affixing the heart monitor that would diagnose their abnormal predisposing heart rhythm.

In the future, we will be smarter about preventing disease. Genomic information will be readily available to help people judge their risks for hundreds of diseases in advance. This risk profile will be integrated with individualized environmental risk monitoring to help determine where best to target preventive care. Patients will be able to discuss all this with their doctors, and the information will be seamlessly integrated into their medical records. Data sharing will mean that someone with a rare disease in Des Moines, Iowa, will be easily connected to the patient in Darwin, Australia, who just presented with the same disease and a variant in the same gene. In the future, by adjusting for genetic risk at scale, hundreds of millions of people around the world will benefit from more accurate, individualized prediction and prevention. The care you receive from your doctor and the screening she chooses to carry out will be informed by this lifetime genetic risk—a risk tailored not just to your overall ethnicity but to the ancestry of specific segments of your genome (we are all glorious genetic mixtures, after all). Your electronic medical record will automatically use your genetically determined ancestry whenever a new test result is returned so your result can be compared to a normal population that most closely matches you. Need a new medication? Every time a medicine is prescribed, the system will check to make sure you get the right dose of the right drug, according to your genome's estimate of your likelihood to respond. Meanwhile, medical devices will be much more personal. Your smartwatch will start to take account of your genetic risk and tune itself to early detection of the diseases you are most at risk for. Perhaps you have a particularly high genetic and lifestyle risk for the heart rhythm atrial fibrillation, predisposing you to strokes (blood clots in the brain). Your device will account for that in how sensitive it is in diagnosing that rhythm. Perhaps you have a high genetic risk for Parkinson's disease. Your watch will take this into account as it analyzes your walking patterns every time you take a step.

This might all sound futuristic, but just think back to 2009 where our book began. There were only a handful of individuals in the world who had ever had their genomes sequenced. By the time this book is published, that number will be in the millions. Indeed, I recently attended an international meeting where more than thirty-six countries with genome programs were represented. Many of these programs were individually making plans to sequence the genomes of millions in their own countries.

And we won't just be sequencing human genomes. As we mentioned earlier, we will be sequencing pathogens too. Our lives are, in so many ways, shaped by the microorganisms that live on us, inside us, and around us. The estimated forty trillion bacteria in our gut, for example, help us digest everything we eat. Disturbing the healthy bacteria that colonize our gut can lead to dangerous blood infections. Today, to diagnose blood infections like those or lung infections like pneumonia, we take a sample of blood or sputum and grow the accompanying bacteria in a lab under special conditions. Eventually, the bacteria are stained and inspected under the microscope and exposed to various antibiotics to determine which ones can kill them. This process takes days. Meanwhile, we treat the patient with our "best guess" antibiotics. In the future, we will sequence those bacteria immediately and map their DNA to a library of microorganism genomes—giving an answer within hours. And it's not just bacteria, of course. Viruses also cause disease, sometimes on a global scale.

In 2020, the world was gripped by a pandemic caused by the coronavirus SARS-CoV-2 (severe acute respiratory syndrome coronavirus 2), a bat virus that spilled over into the human population somewhere in Hubei Province, China. Its toll dominated the front pages of newspapers for most of the year, along with graphs revealing the trajectory of each country's case counts over time—a measure of their competence in suppressing the infection. With no vaccine yet available, no effective therapy, and no cure, Google searches for "exponential growth" soared, mirroring the exponential spread of the virus itself. The world soberly reflected that the death of even a small percentage of billions of people is a very large number indeed. To flatten the curve of rapidly rising infections, most countries initially chose to lock down—closing schools, shops, and public places, limiting travel, and banning mass gatherings like sporting events, weddings, and funerals. Roads were deserted. Malls were quiet. Entire workplaces switched to online video conferencing from home. The air quality improved. Homemade cloth masks proliferated. Yet health care systems in cities that didn't act fast enough were rapidly overwhelmed as the new virus spread through their concentrated populations. Intensive care units overflowed. Personal protective equipment (PPE) for doctors and nurses ran short. Reports from health care workers on the front lines mirrored historical wartime reports. New York City's hospitals and morgues brought in refrigerated trucks to cope with the unprecedented number of dead bodies. The crisis echoed an earlier, more basic time when bodies felled by the 1918 flu pandemic accumulated faster than they could be buried.

Horrifying as this was, it was entirely predictable. For years, scientists, public health experts, thought leaders like Bill Gates, and even Hollywood directors (via films like *Contagion*) had urged governments to prepare for the inevitable: a virus with just the right infectivity, virulence, and opportunity to cause global disarray.

So while many national governments stood paralyzed with the dawning realization that they had acted too late, the scientific community swung into action, moving with unprecedented speed and a singular common purpose: fight the virus, save the world. Virologists, epidemiologists, critical care and infectious disease doctors, geneticists, sociologists, and disaster management experts all turned their attention to perhaps the first globally acknowledged international threat of the information age. They pushed the science forward with torrents of messages exchanged on Twitter, collaborations across the globe, and scientific papers that were made public before their digital ink was even dry. And the heavy artillery in this fight? Genomics.

Within weeks of the first case reports from Wuhan in late December 2019, the genome of the culprit virus, totaling thirty thousand letters of

RNA, had been fully sequenced by Chinese scientists. That sequence identified it as a novel coronavirus: a class of virus first described by Scottish virologist June Almeida, while working at St Thomas' Hospital in London in 1964, who named it, poetically, for the crown-like appearance of the "spike" proteins sticking out from its surface. Since then, coronaviruses had been found to be a cause of the common cold. A coronavirus was also the cause of the 2003 SARS epidemic. The rapid availability of the COVID-19 viral sequence meant that genetic tests could now be developed to determine who was actively infected. Labs from around the world immediately set about designing tests to detect the new virus, with many going from design to completion in just a few weeks. Some tests developed by the early summer of 2020 took as little as thirty minutes to complete, and—like an over-the-counter pregnancy test—gave a simple yes-or-no visual readout. With coordinated government-level planning, such tests could have been available to millions of people just a few weeks after the virus was first sequenced. Coupled with a clear system for tracing each infected person's contacts, this could have saved hundreds of thousands of lives, not to mention the livelihoods of millions around the world. It could have insulated us from a global pandemic that we had the scientific knowledge, but not the political will, to control.

Meanwhile, sequencing the SARS-CoV-2 genome gave scientists another remarkable ability: to reveal the story of how the virus spread across the globe. As any virus copies itself, over and over, within its human hosts, it will inevitably accumulate small changes to its genome. By comparing virus genomes retrieved from the nasal cavities of infected patients around the world, scientists can create a genetic map of pandemic history unfolding in real time. This is how we learned, for example, that most infections on the West Coast of the United States were seeded by infected travelers arriving directly from China, while outbreaks on the East Coast mostly involved infected travelers coming by way of Europe.

Perhaps the most important consequence of obtaining the viral genome was that with unprecedented speed, multiple vaccine candidates were advanced into clinical trials. While some vaccines, like those from the Chinese company Sinovac, were based on the traditional approach of inactivating a virus by exposing it to heat or formaldehyde (both of which stop the virus from multiplying but leave its proteins intact to stimulate our immune systems), others were based on newer, primarily genomic approaches, where parts of the virus genome are delivered for our human cells to make the virus protein. A good example of this genomic approach was a vaccine developed by Adrian Hill and Sarah Gilbert at Oxford University's Jenner Institute, and produced by the British pharmaceutical company AstraZeneca. It involved placing parts of the coronavirus genome into the "shell" of a nonreplicating chimpanzee virus. This Trojan horse co-opts your own human cells to produce the viral protein. In a related technique that avoids the use of any virus, the Boston-based biotechnology company Moderna used tiny lipid particles to deliver an RNA message coding for coronavirus proteins. Although these genomic approaches were not new, the pandemic rapidly accelerated their development into clinical trials. The genomic age of vaccines had begun.

Could even more widespread use of genomics have gotten us further ahead of this pandemic to begin with? The answer is yes. Consider these findings from a group of engineers and epidemiologists at Yale: levels of coronavirus RNA in municipal wastewater, they found, begin to spike a full *seven days* before a corresponding uptick in positive COVID-19 tests at the local hospitals. Waste from toilets and drains, in other words, can provide an early-warning system—giving a local community advance notice that new infections are on the rise. The importance of this finding was underscored just days later, when a report from Columbia University suggested that in the early, exponential-spread phase of the pandemic, locking down communities even one day earlier could have saved tens of thousands of lives in the United States alone. There are already robust systems in place for monitoring water quality. Drinking water in the United States, for example, is routinely tested for over ninety different contaminants. In the future, wastewater could similarly be monitored, cheaply and easily, for genomic signatures of thousands of new and long-familiar microorganisms, providing an early-warning system for infectious disease.

So genomics will help us prevent disease at an individual level and at a population level. After we have read the genomes of millions of humans and billions of pathogens, where will we be in terms of our ability to edit those genomes to defeat disease?

As we saw in chapter 19, we recently entered a golden age of genetic therapy. Many approved therapies are coming to market, transforming our ability to treat—or even cure—devastating and fatal illnesses like hemophilia, spinal muscular atrophy, severe combined immunodeficiency, retinal diseases, sickle cell disease, and even some cancers. Even more exciting, most of these advances occurred before the introduction of CRISPR gene editing, which promises to have a major impact on future therapies. We have a realistic chance of reversing many intractable illnesses, with treatments that may need to be given by injection as rarely as once a year. Some, in fact, may only need to be given once (Sek Kathiresan, who you met in chapter 17, is working on a one-shot CRISPR-based vaccination for heart disease). Yet while we are good at delivering genetic therapy to the liver, to the eye, and to bone marrow stem cells outside of the body, we are only beginning to learn how to efficiently deliver gene therapy to our many other organs. The next few years will bring newfound abilities to deliver genetic therapy to our muscles, our lungs, the nervous system, the brain, and the organ that has been my obsession for decades, the heart.

We will likely also be able to use CRISPR to defeat viruses, not least because over millions of years, this is exactly what bacteria have been using CRISPR for: remembering and killing viruses is what CRISPR evolved to do. Harnessing it inside human cells to kill viruses is something Stanley Qi, my colleague at Stanford, showed was possible early in the COVID-19 pandemic. The reach of gene therapy will be far and wide.

Of course, such innovation doesn't come cheap. The multimillion-dollar cost of some gene therapies has produced heated debate as to whether these prices are fair, who should pay, and in turn, how drug companies should recoup the considerable investment required to bring such drugs safely to market. Should individuals or society pay? Will the capital markets support this model? Should the cost be related to the potential savings to the health care system? These and many other questions are in our near-term future, and we will need thoughtful ethicists and lawmakers to help our society come to fair conclusions. Just because we can, doesn't mean we should. Nor does it mean that we can afford it.

If you had told me what my life and career would hold as I was growing up in Scotland in the 1970s and '80s, I wouldn't have believed you. The idea that my preteen computer skills, sharpened while trying to impress my friends with racehorse games, might one day be useful for crunching genomes would have seemed preposterous and far-fetched. In those pre-Apple, pre-Microsoft, pre-Google, pre-internet days, simply word processing on a computer screen seemed futuristic. Editing a genome? Absurd. Back then, all I knew was that I would try to find some path toward helping people afflicted by disease. By becoming a doctor, I hoped that I would have the chance to care for people, in the best way I could, wherever I ended up.

In these pages, I have tried to explain what enormous privilege I feel getting up every morning and, in direct, or often extremely indirect, ways trying to make people's lives better. To share that effort with remarkable friends and colleagues in the extraordinary environment of Stanford is something I am grateful for every day. While I have highlighted our own efforts in this book, genomic medicine has moved forward as a result of efforts from many groups around the world, each helping the other, the next building on the work of the last. I am inspired and humbled by the brilliance and creativity of my colleagues.

The final words are for my patients. I stand in awe of them. They are the reason I get up in the morning. They make me laugh. They make me cry. Sometimes, they berate me. Sometimes, they hug me. I tell all my new patients that they are now part of our extended family. We walk beside them as they go through life's ups and downs, but we never forget that when we move on to the next patient, they are still there, living with the disease and handling the havoc it wreaks. The people whose stories I have told in these pages inspire me to do more—to do better—in pursuit of understanding the genome ever more deeply, to enable us to treat disease ever more precisely. These are still, after all, the early days of our genome odyssey. Personally, I can't wait to turn the page and find out what happens next.

# Acknowledgments

It seems apt to characterize my writing experience for this book as its own *adventurous journey* and, just as Odysseus couldn't have survived without a lot of help from celestial powers, neither could this book have come together without the help and inspiration of many.

First and foremost, I would like to thank my patients and their families. Their generosity in spending time with me beyond the clinic and their selflessness in letting me tell their stories is humbling. I am thankful every day for the privilege of being a small part of their lives. Since this book went to press, Bertrand Might's remarkable mortal journey came to a premature end. His courage and strength, his purity and joy have been an inspiration and his memory will live on for decades to come.

To the scientists, collaborators, and friends who have generously shared their experiences and delved with me into our collective memories, I can't thank you enough. Every colleague mentioned in this book spent time helping me craft these pages and correcting my mistakes (the responsibility for any remaining errors is entirely mine). A special thanks goes to three people who have shared more days and more patients with me than any other: Matthew Wheeler, Heidi Salisbury and Colleen Caleshu. For fellow genome explorers surveying uncharted territory, I could not have asked for better companions. Many other colleagues and friends were kind enough to discuss sections of the book with me and particular thanks goes to those who read the entire manuscript: Josh Knowles, Marco Perez, Vicki Parikh, Mikael Mattson, Steve Quake, Megan Grove, Colleen Caleshu, James Priest, Roger Burnell, and Susan Schwartzwald. Les Biesecker gave such good feedback on the chapter in which he appeared that I *wished* I had sent him the whole book. Francis Collins was extremely kind in providing timely feedback despite having the minor distraction of running the NIH. A huge thank-you also goes to my Stanford lab group, not least for their hard work and dedication, but also for truly excellent book title suggestions.

A special thanks goes to Kyla Dunn, whom you met in chapters 12 and 13. Suffice it to say, when a Peabody and Emmy Award–winning writer and producer turned genetic counselor indicates she might be open to giving feedback on your genome book, there is only one right answer. I am in awe of Kyla's natural talent for crafting narrative and turning phrases. I learned so much from her mastery of the art of words and am eternally grateful for the time she spent thinking about how to make my words better.

A singular shout-out goes to two people in my group who do not appear as characters in the book but without whom neither our clinical work nor our laboratory science would operate at all: Brooke Zelnik and Terra Coakley. Terra and Brooke have each at different times taken on the onerous task of managing my calendar and for that alone they deserve thanks. But they deserve gratitude for so much more that they do: they are the fulcrum around which our whole enterprise turns. From weekly meetings to annual events, small contracts to multimillion-dollar grants, from small social gatherings to fundraising galas, they take it all in stride, organizing students, CEOs, and Nobel Prize winners with the same grace and happy smiles. Whether I am lost on a rainy Stockholm street, aimlessly wandering a closed Shanghai airport, or sitting on a curiously empty Zoom call, I really couldn't survive without them.

My academic mentors taught me so much and continue to do so: Nanette Mutrie and Neil Spurway in Glasgow; Barbara Casadei, Hugh Watkins, Stefan Neubauer, and John Bell in Oxford; Vic Froelicher, Tom Quertermous, and Randy Vagelos at Stanford. And to the Stanford leaders—Alan Yeung, Bob Harrington, Lloyd Minor, and David Entwistle—thanks for your belief in me.

A special word goes to Abraham Verghese, whose writing I have admired for years. It was Abraham to whom I turned at the moment I realized I might be writing a book. For introducing me to our now shared agent, the amazing Mary Evans, I cannot thank him enough. I smile whenever I see Mary's name on my cell phone because I know the ensuing conversation will be stimulating, entertaining, and informative in equal measure. Mary, I am so grateful to have your help in navigating the unfamiliar world of book publishing. A better advocate no author could hope for.

I am most grateful to Mary for introducing me to the incredible team at Celadon Books. I feel so lucky, as a first-time author, to have a team that seems at once to bring the benefits of a large publishing house while at the same time offering the high touch, individualized care, and nurture of a boutique operation. I am so appreciative of the discerning, insightful, and delicately delivered editorial wisdom of Jamie Raab. Drawing on decades of experience but with a love of story and delight in detail that seems only heightened by those years, Jamie's editorial pen was my personal mentor in learning this craft. For her patience with my learning curve and for her consummate art, I can't thank her enough. In a similar way, Randi Kramer's keen eye greatly improved so many pages of this book, I lost count. To Jamie, Randi, and the other members of the Celadon team, I offer such grateful thanks.

Finally, I want to thank my family. My mum and dad inspired me to go into medicine: they bought me my first book, my first computer, and my first stethoscope. Every day, I strive to offer the care and compassion to my patients I learned watching them care for theirs. My brother, Rod, and sister, Doranne, understand me in ways only siblings can. They were kind enough to read early versions of chapters and provide unfiltered feedback. I am thankful beyond words for the health and happiness of my own children: Katryn, Fraser, and Cameron. They keep me sane and occasionally drive me crazy. I hope in a small way to make the world a better place for them: they are my everything. The last words I have saved for my wife. I think Fiona always knew I would one day try to write a book in my "spare" time. I think she didn't necessarily realize it would be while we both had full-time jobs and were in the throes of bringing up three young children. And so, Fiona, for supporting me through medical school, a residency, a Ph.D., a post-doc, and a cardiology fellowship; for the late nights in the lab and the middle-of-the-night rude awakenings from my pager; for saying "yes" to a crazy California adventure; for the weekends and weekdays when I buried my head in my computer to work on this book; for listening to my stories about Buffalo and Sherlock Holmes and, despite that, actually reading the book and providing great feedback; for keeping my feet firmly on the ground; and for being my partner in life's great adventure, I love you and I can never thank you enough.

# Notes

## Preface

ix *the first genome of a tiny organism was decoded* Discussed in a few chapters of the book, the first organism to have its genome sequenced was PhiX174, a so-called bacteriophage, a virus that infects bacteria. It was first sequenced by Frederick Sanger. Sanger F, Air GM, Barrell BG, et al. Nucleotide sequence of bacteriophage phi X174 DNA. *Nature.* 1977;265(5596):687-695.

ix *not even your identical twin* Contrary to a popularly held view, identical twins do not have identical genomes. This is for a variety of reasons to do with genetic variations that arise after the embryo is formed. Also, chemical changes that occur through life change the way the genome is activated, and these changes are unique to each individual. Bruder CEG, Piotrowski A, Gijsbers AACJ, et al. Phenotypically concordant and discordant monozygotic twins display different DNA copy-number-variation profiles. *Am J Hum Genet.* 2008;82(3):763-771; Lyu G, Zhang C, Ling T, et al. Genome and epigenome analysis of monozygotic twins discordant for congenital heart disease. *BMC Genomics.* 2018;19(1):428; Do Identical Twins Have the Same DNA? BioTechniques. https://www.biotechniques.com/omics/not-so-identical-twins/. Published November 26, 2018. Accessed March 29, 2020.

x *right-handed double helix* A right-handed helix turns in a clockwise direction, which is to say, if you look down "through the helix," it appears to turn away from you in a clockwise direction. Scientists sometimes do not inform illustrators that this matters, and so images of DNA with a left-handed helix are scattered around the internet. This happened to me once when a PR firm contracted by Stanford generated art for a marketing campaign with a left-handed helix (we quickly corrected the error).

x *could stretch to the moon and back thousands of times* Calculations of the length of DNA per cell are drawn from the number of bases multiplied by the distance between bases. There remains significant debate about these kinds of numbers, and it is usually not found in the scientific literature. I use estimates here from: Length of Uncoiled Human DNA. Skeptics Stack Exchange. https://skeptics.stackexchange.com/questions/10606/length-of-uncoiled-human-dna. Accessed January 26, 2020; Crew B. Here's How Many Cells in Your Body Aren't Actually Human. ScienceAlert. https://www.sciencealert.com/how-many-bacteria-cells-outnumber-human-cells-microbiome-science. Accessed January 31, 2020; Yong E. *I*

*Contain Multitudes: The Microbes Within Us and a Grander View of Life.* New York: Random House; 2016.

xii *Richard Dawkins's* The Selfish Gene Dawkins R. *The Selfish Gene.* Oxford, UK: Oxford University Press; 1976.

## PART I: THE EARLY GENOMES

### 1. Patient Zero

3 *His mom recalled* Audio interview with Lynn Bellomi, February 2, 2020.

4 FOXG1 *syndrome* FOXG1 syndrome. Genetics Home Reference. https://ghr.nlm.nih.gov/condition/foxg1-syndrome. Accessed March 29, 2020.

6 *Trait-o-matic* Trait-o-matic was developed by Xiaodi Wu and Alexander "Sasha" Wait-Zaranek from George Church's group. The work was part of the Harvard Personal Genome Project: The Harvard Personal Genome Project. https://pgp.med.harvard.edu/. Accessed March 29, 2020.

6 *The Human Genome Project had been funded for $3 billion* The Human Genome Project funding is described in various places. I used these estimates: Genomics. Energy.gov. https://www.energy.gov/science/initiatives/genomics. Accessed March 29, 2020; Watson JD, Jordan E. The Human Genome Program at the National Institutes of Health. *Genomics.* 1989;5(3):654-656.

7 *a race to be the first to sequence a human genome* The Human Genome Project paper was published alongside the private project. Lander ES, Linton LM, Birren B, et al. Initial sequencing and analysis of the human genome. *Nature.* 2001;409(6822):860-921; Venter JC, Adams MD, Myers EW, et al. The sequence of the human genome. *Science.* 2001;291(5507):1304-1351.

7 *An anonymous Han Chinese man* Sequencing of an Asian individual: Wang J, Wang W, Li R, et al. The diploid genome sequence of an Asian individual. *Nature.* 2008;456(7218):60-65.

7 *had his genome sequenced* Sequencing of James Watson: Wheeler DA, Srinivasan M, Egholm M, et al. The complete genome of an individual by massively parallel DNA sequencing. *Nature.* 2008;452(7189):872-876.

7 *Steve sequenced his own genome* Initial publication of Steve Quake's genome sequence: Pushkarev D, Neff NF, Quake SR. Single-molecule sequencing of an individual human genome. *Nat Biotechnol.* 2009;27(9):847-850.

8 *Moore's law* Over 50 Years of Moore's Law. Intel. https://www.intel.com/content/www/us/en/silicon-innovations/moores-law-technology.html. Accessed March 29, 2020; Moore's Law. Computer History Museum. https://www.computerhistory.org/revolution/digital-logic/12/267. Accessed March 29, 2020.

8 *by releasing a graph* The Cost of Sequencing a Human Genome. Genome.gov. https://www.genome.gov/about-genomics/fact-sheets/Sequencing-Human-Genome-cost. Accessed March 29, 2020.

13 *These vary enormously in size* Largest and smallest genes taken from this textbook: Strachan T, Read AP. *Human Molecular Genetics.* New York: Garland; 2018. doi:10.1201/9780429448362. Some other genome anatomy facts from: Platzer M. The human genome and its upcoming dynamics. *Genome Dyn.* 2006;2:1-16.

14 *invented by Frederick Sanger* Frederick Sanger's biographical details from: Berg P. Fred Sanger: A memorial tribute. *Proc Natl Acad Sci USA.* 2014;111(3):883-884.

14 *To understand Sanger sequencing* References for Sanger and next-generation sequencing: Heather JM, Chain B. The sequence of sequencers: The history of sequencing DNA. *Genomics.* 2016;107(1):1-8; Goodwin S, McPherson JD, McCombie WR. Coming of age: Ten years of next-generation sequencing technologies. *Nat Rev Genet.* 2016;17(6):333-351.

Another technology was invented around the same time as Sanger sequencing by Walter "Wally" Gilbert. Gilbert was a physicist turned biochemist at Harvard who worked closely with James Watson for many years. His technique involved chemical modification and cutting of DNA, but also used extensive radioactivity and so, despite initially surpassing Sanger's technique in popularity, it was soon supplanted by the improvements in Sanger's technique.

16 *before another individual's genome was published* Papers for the first few genomes included estimates of cost and time it took: Lander ES, Linton LM, Birren B, et al. Initial sequencing and analysis of the human genome. *Nature.* 2001;409(6822):860-921; Venter JC, Adams MD, Myers EW, et al. The sequence of the human genome. *Science.* 2001;291(5507):1304-1351; Wang J, Wang W, Li R, et al. The diploid genome sequence of an Asian individual. *Nature.* 2008;456(7218):60-65; Wheeler DA, Srinivasan M, Egholm M, et al. The complete genome of an individual by massively parallel DNA sequencing. *Nature.* 2008;452(7189):872-876; Bentley DR, Balasubramanian S, Swerdlow HP, et al. Accurate whole human genome sequencing using reversible terminator chemistry. *Nature.* 2008;456(7218):53-59; Kim J-I, Ju YS, Park H, et al. A highly annotated whole-genome sequence of a Korean individual. *Nature.* 2009;460(7258):1011-1015.

16 *a company called 454* The numbers 454 represented the code name by which the technology was known when it was first invented and their significance has never been properly explained, at least not in the public domain.

16 *next-generation technologies have in common* One of the first next generation approaches was called *Polony* sequencing and emerged from George Church's lab at Harvard. Shendure J, Porreca GJ, Reppas NB, et al. Accurate multiplex polony sequencing of an evolved bacterial genome. *Science.* 2005;309(5741):1728-1732.

Polony sequencing was spearheaded by Jay Shendure and Greg Porreca building on the work of Rob Mitra. See references at: Open Source Next Generation Sequencing Technology. Harvard Molecular Technologies. http://arep.med.harvard.edu/Polonator/. Accessed December 28, 2016.

Deriving its name from the *polymerase* of DNA polymerase and *colonies* from the principle of reading DNA sequences from millions of molecules, each one amplified within its own tiny water droplet in an emulsion of oil (a colony of identical DNA molecules). Jay Shendure has subsequently developed a vast array of genomic technologies; in particular, he was among the first to apply exome sequencing to patients (four patients with the same genetic syndrome) in one of a series of collaborations with pioneer Deborah "Debbie" Nickerson. Ng SB, Turner EH, Robertson PD, et al. Targeted capture and massively parallel sequencing of 12 human exomes. *Nature.* 2009;461(7261):272-276. Another early pioneer of exome sequencing was Richard "Rick" Lifton: Genetic diagnosis by whole exome capture and massively parallel DNA sequencing Proc Natl Acad Sci U S A. 2009 Nov 10; 106(45): 19096-19101

## 2. Team of Teams

22 *PharmGKB* PharmGKB. http://www.pharmgkb.org.

23 *genetic counseling* The history of genetic counseling in more detail can be found here: Stern AM. *Telling Genes: The Story of Genetic Counseling in America.* Baltimore: JHU Press; 2012.

24 *incidentalome* The word *incidentalome* was coined in this editorial: Kohane IS, Masys DR, Altman RB. The incidentalome: a threat to genomic medicine. *JAMA.* 2006;296(2):212-215.

26 *René Laennec* Roguin A. Rene Theophile Hyacinthe Laënnec (1781–1826): The man behind the stethoscope. *Clin Med Res.* 2006;4(3):230-235.

26 *Augustus Waller* Augustus Desire Waller biography from the *Oxford Dictionary of National Biography*: Waller AD. A Demonstration on Man of Electromotive Changes accompanying the Heart's Beat. *J Physiol.* 1887;8(5):229-234; *Oxford Dictionary of National Biography.* Oxford, UK: Oxford University Press.

26 *used to demonstrate* Houses of Parliament conversation on Waller and Jimmie: Royal Society Conversazione (Public Experiment on Bulldog). Hansard. http://hansard.millbanksystems.com/commons/1909/jul/08/royal-society-conversazione-public. Accessed December 30, 2016.

27 *slowly* brrrred *out of the machine* For some reason, this always reminds me of Saturday afternoons watching TV growing up. The BBC used to use a teleprinter to reveal football (soccer) results letter by letter on the screen, which would be voiced over by a commentator like David Coleman. https://www.youtube.com/watch?v=-_V43QT7mrg&feature=youtu.be&t=20s. Accessed August 8, 2020.

This clip includes the phrase, "Motherwell, the bottom club, have only won two matches all season." Motherwell was my team, and this kind of win-loss record prepared me for a lifetime of disappointment in the Scottish national football team.

28 *pushes the water away* The science of water in oil emulsions: Stability of Oil Emulsions. PetroWiki. http://petrowiki.org/Stability_of_oil_emulsions.

29 *detergent* Handbook of detergents Part D Formulation: Showell M. Part D: Formulation. In *Handbook of Detergents.* Boca Raton, FL: CRC Press; 2016.

29 *Lp(a)—pronounced "L P little a"* The paper by the Oxford group describing the association of single nucleotide variants with Lp(a) levels and coronary risk: Clarke R, Peden JF, Hopewell JC, et al. Genetic variants associated with Lp(a) lipoprotein level and coronary disease. *N Engl J Med.* 2009;361(26):2518-2528.

33 *how comprehensive these guidelines were* These guidelines were commonly referred to as *ATPIII,* which is slightly shorter than the *Third Report of the National Cholesterol Education Program Expert Panel on Detection, Evaluation, and Treatment of High Blood Cholesterol in Adults (Adult Treatment Panel III).* NCEP ATP-III Cholesterol Guidelines. ScyMed. http://www.scymed.com/en/smnxdj/edzr/edzr9610.htm. Accessed December 30, 2016.

35 *one of the editors* Journal editors are the unsung heroes of academic publishing. A good editor will not only help define the character of a journal, arbitrate among referees, and calm overanxious authors but will also, most importantly, improve the papers. In the open access movement of publishing, there has been a push to "disintermediate" the publisher. Certainly, academic publishing is highly lucrative for executives and shareholders of large companies and takes advantage of free labor in the form of scientists producing, then reviewing, and even editing content to which the publisher then takes ownership. But it would be a great loss if, in the move to open access, we were to disintermediate the editors. A good editor can transform a lousy paper and elevate a great one.

36 *The papers appeared* The papers were published in *The Lancet* on May 1, 2010: Ashley EA, Butte AJ, Wheeler MT, et al. Clinical assessment incorporating a personal genome. *Lancet.* 2010;375(9725):1525-1535; Ormond KE, Wheeler MT, Hudgins L, et al. Challenges in the clinical application of whole-genome sequencing. *Lancet.* 2010;375(9727):1749-1751; Samani NJ, Tomaszewski M, Schunkert H. The personal genome—the future of personalised medicine? *Lancet.* 2010;375(9725):1497-1498.

37 *"doctor-patient" interview* The NPR interview is found here: Knox R. Genome Seen As Medical Crystal Ball. NPR. https://www.npr.org/templates/story/story.php?storyId=126396839. Published April 30, 2010. Accessed April 7, 2020.

37 *we were celebrities* A representative selection of the news articles: Marcus AD. How Genetic Testing May Spot Disease Risk. *Wall Street Journal.* https://www.wsj.com/articles/SB10001424052748704342604575222082732063418. Published May 4, 2010. Accessed April 7, 2020; Krieger LM. Stanford Bioengineer Explores Own Genome. *Mercury News.* https://www.mercurynews.com/2010/04/29/stanford-bioengineer-explores-own-genome/. Published April 29, 2010. Accessed April 7, 2020; Nainggolan L. First Clinical Interpretation of an Entire Human Genome "Exemplar." Medscape. https://www.medscape.com/viewarticle/721083. Published April 30, 2010. Accessed April 7, 2020; Fox M. Gene Scan Shows Man's Risk for Heart Attack, Cancer. Reuters. https://www.reuters.com/article/us-genes-disease

-idUSTRE63S62J20100429. Published April 29, 2010. Accessed April 7, 2020; Sample I. Healthy Genome Used to Predict Disease Risk in Later Life. *Guardian.* http://www.theguardian.com/science/2010/apr/29/healthy-genome-predict-disease-risk. Published April 29, 2010. Accessed April 7, 2020.

## 3. Once Removed

39 *Rich told me later* Details of Richie's case come from email correspondence and personal conversations at the time and over the years since with his mom and dad. The quotes come from an in-person interview conducted with Rich Quake and his daughter on May 3, 2019.

41 *most common autopsy finding* A variety of papers have reported this. We summarized the findings here: Ullal AJ, Abdelfattah RS, Ashley EA, Froelicher VF. Hypertrophic cardiomyopathy as a cause of sudden cardiac death in the young: A meta-analysis. *Am J Med.* January 2016. doi:10.1016/j.amjmed.2015.12.027.

Jonathan Drezner summarizes a decade of data from the NCAA: Harmon KG, Asif IM, Maleszewski JJ, et al. Incidence, cause, and comparative frequency of sudden cardiac death in national collegiate athletic association athletes: A decade in review. *Circulation.* 2015;132(1):10-19.

45 *Rick presented our results* Rick presented our findings at the American College of Cardiology meeting: Dewey FE, Wheeler MT, Cordero S, et al. Molecular autopsy for sudden cardiac death using whole genome sequencing. *J Am Coll Cardiol.* 2011;57(14, Supplement):E1159.

45 *wrote a beautiful piece* Krista Conger's piece can be found here: Conger K. The Genome Is Out of the Bag. *Stanford Medicine.* http://sm.stanford.edu/archive/stanmed/2010fall/article1.html. Accessed April 5, 2020.

45 *identified the molecular cause of an electrical disease of the heart* Michael Ackerman's case report: Ackerman MJ, Tester DJ, Porter C-BJ, Edwards WD. Molecular diagnosis of the inherited long-QT syndrome in a woman who died after near-drowning. *N Engl J Med.* 1999;341(15):1121-1125. doi:10.1056/nejm199910073411504.

45 *a group led by Chris Semsarian* Study from Christopher Semsarian and Jon Skinner: Bagnall RD, Weintraub RG, Ingles J, et al. A prospective study of sudden cardiac death among children and young adults. *N Engl J Med.* 2016;374(25):2441-2452.

47 *put it better* We have had the pleasure of working with DJ Patil on data science related projects. I interviewed him for our Big Data in Biomedicine conference at Stanford in 2019: Stanford Medicine. DJ Patil, Devoted-2019 Stanford Medicine Big Data | Precision Health.mp4. https://www.youtube.com/watch?v=mK3N7xQb_mw. Published July 3, 2019. Accessed April 5, 2020.

A photo of this quote written on White House notepaper can be found on social media. https://twitter.com/dpatil/status/1093569468880416768

## 4. Genome Illumination

48 *had its origins* The story of the early origins of Solexa comes from in-person conversations with Clive Brown, John West, and Kevin Davies. Some details are from articles written by Kevin Davies: Davies K. 13 Years Ago, a Beer Summit in an English Pub Led to the Birth of Solexa and—for Now at Least—the World's Most Popular Second-Generation Sequencing Technology. Bio-IT World. http://www.bio-itworld.com/2010/issues/sept-oct/solexa.html. Accessed January 11, 2019.

And some are from his excellent book *The $1,000 Genome:* Davies K. *The $1,000 Genome: The Scientific Breakthrough That Will Change Our Lives.* New York: Free Press; 2010.

49 *named Solexa* Nick McCooke used to say the name *Solexa* was a random name chosen from a list of "possible biotech names" generated by a computer program, cited by Clive Brown.

49 *Enter Clive Brown* In-person interview with Clive Brown November 27, 2018, Stanford campus.

50 *John West* In-person interviews with John West were carried out at the offices of Personalis, Menlo Park, California, on December 4, 2017, and on November 2, 2018. Other biographical details from: John West. Personalis. https://www.personalis.com/john-west/. Published August 17, 2017. Accessed July 10, 2018.

51 *company called Manteia* As reported in: Kitchens F. Lynx and Solexa Buy DNA Cluster Technology from Manteia. GenomeWeb. https://www.genomeweb.com/archive/lynx-and-solexa-buy-dna-cluster-technology-manteia. Published March 25, 2004. Accessed July 6, 2018.

53 *Jay Flatley* Jay Flatley details come from this interview: The DNA Day interview: Jay Flatley, Executive Chairman of Illumina. Helix Blog. https://blog.helix.com/jay-flatley-interview/. Published April 25, 2018. Accessed January 12, 2019.

54 *the second-generation Genome Analyzer was launched* Genome analyzer II output stats: Genome Analyzer IIx System. Illumina. https://www.illumina.com/Documents/products/specifications/specification_genome_analyzer.pdf.

54 *HiSeq* HiSeq 1000 and 2000 output stats: HiSeq Sequencing Systems. Illumina. https://www.illumina.com/documents/products/datasheets/datasheet_hiseq_systems.pdf.

55 *Illumina decided to offer "personal" genome sequencing* Details of the Individualized Genome Service from Illumina come from this interview with Jay Flatley: Davies K. Jay Talking Personal Genomes. Bio-IT World. http://www.bio-itworld.com/2010/issues/sept-oct/flatley.html. Published September 28, 2010. Accessed January 12, 2019.

55 *four people all with the same last name* Illumina announced the sequencing of the West family in 2010: Illumina Announces Its First Full Coverage DNA Sequencing of a Named Family. *BusinessWire.* https://www.businesswire.com/news/home/20100416006128/en/Illumina-Announces-Full-Coverage-DNA-Sequencing-Named. Published April 16, 2010. Accessed April 6, 2020.

## 5. First Family

57 *John and his daughter, Anne The Wall Street Journal*'s Amy Marcus covered Anne West's analysis of their genomes. Marcus AD. Obsessed with Genes (Not Jeans), This Teen Analyzes Family DNA. *Wall Street Journal.* https://www.wsj.com/articles/SB10001424052748704814204575508064149859510. Published October 1, 2010. Accessed April 6, 2020.

59 *Ronald Davis* Ron Davis was recognized along with Elon Musk, Jeff Bezos, Vint Cerf, and others as one of the inventors most likely to be noted by tomorrow's historians as the great inventors of today. Allan N. Who Will Tomorrow's Historians Consider Today's Greatest Inventors? *Atlantic.* October 2013. http://www.theatlantic.com/magazine/archive/2013/11/the-inventors/309534/. Accessed January 2, 2018.

61 *Factor V Leiden* The original Factor V Leiden paper was published in *Nature:* Bertina RM, Koeleman BP, Koster T, et al. Mutation in blood coagulation factor V associated with resistance to activated protein C. *Nature.* 1994;369(6475):64-67.

In the years that followed, its effect size and prevalence was clarified: Miñano A, Ordóñez A, España F, et al. AB0 blood group and risk of venous or arterial thrombosis in carriers of factor V Leiden or prothrombin G20210A polymorphisms. *Haematologica.* 2008;93(5):729-734; Bauer KA. The thrombophilias: Well-defined risk factors with uncertain therapeutic implications. *Ann Intern Med.* 2001;135(5):367-373; Herrmann FH, Koesling M, Schröder W, et al. Prevalence of factor V Leiden mutation in various populations. *Genet Epidemiol.* 1997;14(4):403-411.

## 6. Buffalo Buffalo Buffalo

62 *chicken wings* Calvin Trillin laid out the origin story of the "chicken wing as delicacy" in: Trillin C. An Attempt to Compile a Short History of the Buffalo Chicken Wing. In *The Tummy Trilogy.* New York: Farrar, Straus & Giroux; 1994. 268–275.

As he reports, the least controversial seat of invention for the wing we know today is the Anchor Bar in Buffalo, New York, where, according to legend, the son of the owners arrived home late one night with some friends and the munchies. His mom, Teressa Bellissimo, being Italian, existed in a state of constant readiness to feed her son. Late Friday night being no exception, she fired up the deep fryer, grabbed the leftover wings that no one wanted, cut them in half, deep-fried them, paired them with hot sauce, and served them with celery and blue cheese dressing. The Buffalo chicken wing was born! A decade later, a National Chicken Wing Day was declared (July 29, 1977), and the rest, including Hooters, is history. Cementing its place as the fuel for the genomics revolution, and for reasons that escape me, the leaders of the world-renowned Australian inherited cardiovasular disease group, Jodie Ingles and Chris Semsarian, head straight for Buffalo chicken wings every time they touch down on US soil.

62 *buffalo buffalo buffalo* Buffalo is also a verb (to *buffalo* means "to intimidate"), meaning it is possible to create a grammatically correct sentence using the same word repeated any number of times. To wit: buffalo buffalo. The animals known as buffalo, at times, intimidate. And if those buffalo are intimidating their neighbors: buffalo buffalo buffalo. If those buffalo actually came from Buffalo, New York, then: Buffalo buffalo buffalo Buffalo buffalo. I first enjoyed this wordplay in chapter 7 of Steven Pinker's book: Pinker S. *The Language Instinct: How the Mind Creates Language.* London: Penguin UK; 2003.

William Rapaport claims some degree of "invention" in his blog: Rapaport W. Buffalo Buffalo Buffalo Buffalo Buffalo. University at Buffalo. https://www.cse.buffalo.edu//~rapaport/buffalobuffalo.html. Accessed December 23, 2017.

The earliest known example appeared in the original manuscript for a book by Dmitri Borgmann in 1965 (*Language on Vacation*), though the chapter containing this was not included in the final published version. However, Borgmann included this in his 1967 book, *Beyond Language: Adventures in Word and Thought.* Final thought from a biologist. There are no *actual* buffalo in Buffalo. Although they are often *called* buffalo, North American "buffalo" are, in fact, bison. To be precise, they are *Bison bison* (or even, *Bison bison bison,* but let's not get started on that). Hedrick PW. Conservation genetics and North American bison (Bison bison). *J Hered.* 2009;100(4):411-420.

This is confirmed by, for me, the ultimate authority on the subject, Buffalo Zoo, who note on their website that their "buffalo" are, in fact, bison. So those intimidating buffalo could be located in many places in the world. Just not in Buffalo, New York.

62 *The advertisement was placed* Pieter de Jong was kind enough to share the article and advertisement he placed in *The Buffalo News.* Information on the Human Genome project comes from email and personal conversations with Deanna Church and Pieter de Jong (phone interview with Pieter de Jong, October 30, 2018). Some information derives from the key publication describing the BAC library RPCI-11: Osoegawa K, Mammoser AG, Wu C, et al. A bacterial artificial chromosome library for sequencing the complete human genome. *Genome Res.* 2001;11(3):483-496.

63 *real human's genome is diploid* Many plants and crops have multiple copies of their genomes. Some naturally occurring wild strawberries exhibit decaploidy (ten copies). The advantages and disadvantages of polyploidy are discussed in: Comai L. The advantages and disadvantages of being polyploid. *Nat Rev Genet.* 2005;6(11):836-846; The naturally occuring and cultivated polyploidy of strawberries is discussed in several articles: Hummer KE, Nathewet P, Yanagi T. Decaploidy in Fragaria iturupensis (Rosaceae). *Am J Bot.* 2009;96(3):713-716; Cheng H, Li J, Zhang H, et al. The complete chloroplast genome sequence of strawberry (Fragaria × ananassa Duch.) and comparison with related species of Rosaceae. *Peer J.* 2017;5:e3919.

Artificial plants with ploidy up to 32-ploid have been successfully cultivated.

65 *containing the DNA of an African American* The fact that RPCI-11 could be inferred to be an African American male with mixed ancestry is known to many geneticists but not often discussed. In an interview with Kevin Davies, Deanna Church mentions conference presentations where this was discussed openly: Davies K. Deanna Church on the Reference Genome

Past, Present, and Future. Bio-IT World. http://www.bio-itworld.com/2013/4/22/church-on-reference-genomes-past-present-future.html. Published April 22, 2013. Accessed December 30, 2017.

67 *who inherited what piece of DNA from whom* We often round the percentage of DNA we share with our first-degree relatives to 50 percent. Recall that except for the singular X and Y chromosomes in men, humans normally have two copies of each chromosome and pass one of those on to each child. However, chromosomes aren't passed on down the generations completely intact. They get a little bit of mixing because of a process called *crossing over.* When sperm or eggs are created, there is an exchange of genetic material between similar pairs of chromosomes, meaning that each chromosome in each sperm or egg comes to contain a unique mixture of the genetic material from the chromosomes that person received from their father and mother. This "unique in the world" chromosome is passed on via the sperm or egg to create the new human. Another consequence of this phenomenon is that although it is true that "on average" we share 50 percent of our genetic material with our brothers and sisters, the actual number can vary substantially. One study found it ranged from 37 to 62 percent: Visscher PM, Medland SE, Ferreira MAR, et al. Assumption-free estimation of heritability from genome-wide identity-by-descent sharing between full siblings. *PLOS Genet.* 2006;2(3):e41.

67 *hidden Markov model* Hidden Markov models have been used in diverse settings, including speech recognition and natural language processing. Ghahramani Z, Jordan MI. Factorial Hidden Markov Models. In Touretzky DS, Mozer MC, Hasselmo ME, eds. *Advances in Neural Information Processing Systems 8.* Cambridge, MA: MIT Press; 1996:472-478.

Around the same time that we were working on this model, a group from Seattle led by David Galas and Leroy "Lee" Hood published genome sequencing data including an HMM from four related individuals. (Lee Hood is a giant in the field of genetic sequencing, a prolific inventor whose discoveries were commercialized by Applied Biosystems, the company whose machines were the workhorses for the Human Genome Project.) The genomes in that paper were produced by a company called Complete Genomics, a start-up founded by Cliff Reid, John Curson, and Radoje "Rade" Drmanac. Rade was the technical lead for Complete Genomics and the inventor of the DNA nanoball technology. It became the foundational technology of Complete Genomics, for a time the only major competitor to Illumina in the genome sequencing space. Complete Genomics was bought by the Chinese company BGI in 2012. The technology resurfaced in headline news in 2020, when the price per genome under certain conditions was claimed to be as low as one hundred dollars. Regalado A. China's BGI Says It Can Sequence a Genome for Just $100. *MIT Technology Review.* https://www.technologyreview.com/2020/02/26/905658/china-bgi-100-dollar-genome/. Published February 26, 2020. Accessed June 14, 2020.

68 *something called* phasing The importance of phasing was discussed in this paper: Tewhey R, Bansal V, Torkamani A, Topol EJ, Schork NJ. The importance of phase information for human genomics. *Nat Rev Genet.* 2011;12(3):215-223.

68 *brand-new variants* De novo mutation rates were reported by us as well as by: Roach JC, Glusman G, Smit AFA, et al. Analysis of genetic inheritance in a family quartet by whole-genome sequencing. *Science.* 2010;328(5978):636-639; Kong A, Frigge ML, Masson G, et al. Rate of de novo mutations and the importance of father's age to disease risk. *Nature.* 2012;488(7412):471.

70 *how databases of genetic variants should be structured* Our classification of genetic variants was described in: Ashley EA, Butte AJ, Wheeler MT, et al. Clinical assessment incorporating a personal genome. *Lancet.* 2010;375(9725):1525-1535.

And developed further in: Dewey FE, Chen R, Cordero SP, et al. Phased whole-genome genetic risk in a family quartet using a major allele reference sequence. *PLOS Genet.* 2011;7(9):e1002280.

A similar schema was also described in: Berg JS, Khoury MJ, Evans JP. Deploying whole genome sequencing in clinical practice and public health: Meeting the challenge one bin at a time. *Genet Med.* 2011;13(6):499-504.

70 *a bone marrow transplant* One of the most famous stories of this early era was that of Nicholas Volker, a young boy with a mystery disease that led to him suffering recurrent bouts of severe abdominal pains requiring over a hundred procedures and surgeries. Doctors were stumped and ready to give up until a team at the Medical College of Wisconsin offered to sequence his exome (the 2 percent of the genome represented by the genes). The team was led by geneticist Howard Jacob, Scottish bioinformatician Elizabeth Worthey, and English pediatrician David Dimmock. Together, they identified Nicholas's condition as an immunodeficiency disorder, not a gastrointestinal disorder, the clear implication being that it could potentially be cured by a bone marrow transplant. After ethical review and counseling, the family opted for the transplant, and Nicholas made a remarkable recovery. The story of his diagnosis was told in a series of Pulitzer Prize–winning articles in the *Milwaukee Journal Sentinel* and in a subsequent book, *One in a Billion.* Herper M. The First Child Saved By DNA Sequencing. *Forbes.* January 2011. https://www.forbes.com/sites/matthewherper/2011/01/05/the-first-child-saved-by-dna-sequencing/. Accessed January 12, 2019; Johnson M, Gallagher K. A Baffling Illness. *Journal Sentinel.* http://archive.jsonline.com/features/health/111641209.html. Published December 18, 2010. Accessed January 12, 2019; Mark Johnson, Kathleen Gallagher, Gary Porter, Lou Saldivar and Alison Sherwood of *Milwaukee Journal Sentinel.* Pulitzer Prizes. https://www.pulitzer.org/winners/mark-johnson-kathleen-gallagher-gary-porter-lou-saldivar-and-alison-sherwood. Accessed January 12, 2019; Johnson M, Gallagher K. *One in a Billion: The Story of Nic Volker and the Dawn of Genomic Medicine.* New York: Simon & Schuster; 2016.

72 *hailed as genome pioneers* The West family analysis was reported in: Marcus AD. Family Pioneers in Exploration of the Genome. *Wall Street Journal.* https://www.wsj.com/articles/SB10001424053111904491704576573022083190718. Published September 16, 2011. Accessed July 16, 2018.

## 7. Starting Up, Reaching Out

74 *Marc Andreessen* The Marc Andreessen quote comes from: Sanghvi R. 17 Quotes from Marc Andreessen & Ron Conway on How To Raise Money. Medium. https://medium.com/how-to-start-a-start-up/17-quotes-from-marc-andreessen-ron-conway-on-how-to-raise-money-d0b710f115f1. Published October 22, 2014. Accessed November 13, 2019.

75 *funkymetals.co.uk* I cofounded this company with my wife in the late 1990s in the early days of e-commerce on the internet. I was starting my Ph.D., and as it was ramping up, I spent some late nights coding up a website in a text editor, showcasing my father-in-law's artisan metalwork. Our most popular item was a "phone tree" on which you could place multiple mobile phones while they were charging.

77 *Bayh-Dole Act* Kastenmeier RW. An Act to Amend the Patent and Trademark Laws; 1980. https://www.congress.gov/bill/96th-congress/house-bill/6933. Accessed April 20, 2020.

78 *enabled Stanford to license the so-called gene splicing patents* Leslie Berlin's tremendous history of a pivotal period in Silicon Valley history told from the perspective of seven individuals, including Niels Riemers of Stanford's Office of Technology Licensing. Berlin L. *Troublemakers: How a Generation of Silicon Valley Upstarts Invented the Future.* New York: Simon & Schuster; 2017.

80 *Concinnity* Concinnity definition from: Concinnity. Oxford Dictionaries. https://en.oxforddictionaries.com/definition/concinnity. Accessed November 18, 2018.

83 *Running the pipeline to identify gene variants was routine by this point* For the Quake genomes, our software was developed in-house. For the West genomes, we used a combination of software we built and Illumina software. Starting at that moment and until today, we and others have used a combination of programs developed by Heng Li of the Dana-Farber Cancer

Institute and by a group led by Mark DePristo and Mark Daly out of the Broad Institute. Li H, Durbin R. Fast and accurate short read alignment with Burrows-Wheeler transform. *Bioinformatics.* 2009;25(14):1754-1760; DePristo MA, Banks E, Poplin R, et al. A framework for variation discovery and genotyping using next-generation DNA sequencing data. *Nat Genet.* 2011;43(5):491-498.

## PART II: DISEASE DETECTIVES

### 8. Undiagnosed

91 *BBC's wildlife correspondent David Attenborough* Sir David Frederick Attenborough is perhaps TV's best known natural historian. He has been an inspiration to millions through a decades-long career spent producing award-winning documentaries featuring mesmerizing natural world videography and his trademark narration (examples include: *Life on Earth, The Living Planet,* and *The Blue Planet*).

92 *The beauty and torture of a traditional medical education* I benefited greatly from a traditional medical education at the 570-year-old University of Glasgow, including a rigorous approach to observation in clinical examination. Such skills are underlined by the rite of passage that is the Membership of the Royal College of Physicians examination (where traditionally, the pass rate was set at the point where 75 percent of the entire class would fail). The classic textbook for the "short cases" is still on my bookshelf: Ryder REJ, Mir MA, Freeman EA. *An Aid to the MRCP Short Cases.* Hoboken, NJ: John Wiley & Sons; 2009. And while it now seems to harken a bygone age, its teachings on active observation remain all too relevant in the present technological age.

93 *Through the Keyhole* Loyd Grossman was the host of this show on the BBC along with David Frost. For sixteen years, he would tour the homes of celebrities describing them in his mid-Atlantic accent. Loyd Grossman. Wikipedia. https://en.wikipedia.org/w/index.php?title=Loyd_Grossman&oldid=946370437. Published March 19, 2020. Accessed May 2, 2020; Through the Keyhole. Wikipedia. https://en.wikipedia.org/w/index.php?title=Through_the_Keyhole&oldid=953132453. Published April 25, 2020. Accessed May 2, 2020; jflitter. The Best of Through the Keyhole (Yorkshire Television) DAMAGED TAPE-August 1988. https://www.youtube.com/watch?v=WhIzhVBQOUY. Published March 23, 2018. Accessed May 2, 2020; Stecklow S. The Wall Street Journal on Americans Adopting British Accents. *Guardian.* http://www.theguardian.com/world/2003/oct/04/usa.theeditorpressreview. Published October 4, 2003. Accessed May 2, 2020.

93 *Arthur Ignatius Conan Doyle* Excerpts from Conan Doyle's writing are taken from: Doyle AC. *The Complete Sherlock Holmes: All 56 Stories & 4 Novels.* New Delhi: General Press; 2016.

93 *leaned heavily on the influence of one of his medical mentors* Medical connections within the stories of Sherlock Holmes have been discussed by several authors: Reed J. A medical perspective on the adventures of Sherlock Holmes. *Med Humanit.* 2001;27(2):76-81; Key JD, Rodin AE. Medical reputation and literary creation: an essay on Arthur Conan Doyle versus Sherlock Holmes 1887-1987. *Adler Mus Bull.* 1987;13(2):21-25.

93 *Joseph Bell* Details from: *Oxford Dictionary of National Biography.* Oxford, UK: Oxford University Press. And from Bell's book: Bell J. A manual of the operations of surgery. Internet Archive. https://archive.org/details/amanualoperatio04bellgoog/page/n7/mode/2up. Accessed May 2, 2020.

94 *Many parallels can be drawn* A small number of publications deal with the kinds of reasoning used by law enforcement in the modern age. A good example is one from: Innes M. *Investigating Murder: Detective Work and the Police Response to Criminal Homicide.* Oxford, UK: Oxford University Press; 2003.

94 *Holmes's process* Maria Konnikova's spectacular book discusses Holmes's process in some detail. (*Mastermind: How to Think Like Sherlock Holmes.* New York: Penguin; 2013) Although Holmes himself refers to his conclusions as "deductions," strictly, it is not deductive reasoning that Holmes is engaged in. Scholars recognize three forms of reasoning: deductive, inductive, and abductive. Deductive reasoning has a long history, as far back as Aristotle (384 B.C.), and is pure logic involving sequential premises leading to a conclusion ("All nerds are fun; John is a nerd; therefore, John is fun"). But deduction tells us no more than the observational statements themselves reveal. Inductive reasoning, in contrast, flows from making observations about the world and then generalizing from these observations ("All swans I've seen to date are white; therefore, all swans are white"). Criminal investigation and diagnostic medicine is based neither on deduction nor induction; it is based on abduction, a form of reasoning that approaches the idea of causality. Abduction involves making observations, using prior knowledge of the world to form a hypothesis as to what caused the events that led to the observations, then testing the hypothesis using specific observational if-then trials. The American philosopher Charles Sanders Peirce writes eloquently about abductive reasoning. Following Peirce, some have argued that crime solving actually involves moving through all stages of inference: observation leads to *abduction* to form a hypothesis from the clues, *induction* to generalize from prior observations of the world, and *deduction* to come to logical conclusions based on the full extension of these premises. Peirce's work is put in context by: Anderson DR. The evolution of Peirce's concept of abduction. *Transactions of the Charles S. Peirce Society.* 1986;22(2):145-164; Burks AW. Peirce's theory of abduction. *Philos Sci.* 1946;13(4):301-306.

94 *sixty-eight diseases, thirty-eight doctors, twenty-two drugs, twelve medical specialties, and three medical journals* Reed J. A medical perspective on the adventures of Sherlock Holmes. *Med Humanit.* 2001;27(2):76-81.

94 *curious incident of the dog in the night* The phrase *the curious incident of the dog in the night time* was used as a title for an acclaimed novel by Mark Haddon: Haddon M. *The Curious Incident of the Dog in the Night-Time.* Washington, D.C.: National Geographic Books; 2007. Later, it was adapted for the stage by Simon Stephens: Haddon M, Stephens S. *The Curious Incident of the Dog in the Night-Time.* doi:10.5040/9781408173381.00000006.

95 *Holmes uses all his senses* The use of smell in diagnosis is reviewed in: Bijland LR, Bomers MK, Smulders YM. Smelling the diagnosis: A review on the use of scent in diagnosing disease. *Neth J Med.* 2013;71(6):300-307.

95 *the Fox TV series* House Details from *House* derived from the Internet Movie Database (IMDb) and Wikipedia: House (TV series). Wikipedia. https://en.wikipedia.org/w/index.php?title=House_(TV_series)&oldid=954022056. Published April 30, 2020. Accessed May 2, 2020. https://paperpile.com/c/zm9uRu/nwdb.

96 *William "Bill" Gahl, MD* Bill Gahl's biography and most details of the early days of the Undiagnosed Diseases Program at the NIH come from a personal interview by teleconference on December 18, 2018. Other details of the origins of the UDP come from personal conversations with Francis Collins, Teri Manolio, and Elizabeth "Betsy" Wilder.

96 The Metabolic Basis of Inherited Disease The first editon is the classic textbook for genetic metabolic disorders. Stanbury JB. *The Metabolic Basis of Inherited Disease.* New York: McGraw-Hill; 1972.

97 *the head of the Office of Rare Diseases Research at the NIH* Stephen Groft was the head of the Office of Rare Disease at the NIH and the person who asked Bill to tackle the undiagnosed patients.

98 *Elias Zerhouni* Biography from personal conversations and this source: Craine A. Elias Zerhouni. *Encyclopædia Britannica.* https://www.britannica.com/biography/Elias-Zerhouni. Accessed May 2, 2020.

101 The Odyssey *The Odyssey* referenced here is the Penguin Classics edition translated by Robert Fagle: Homer. *The Odyssey.* London: Penguin UK; 2003. *The Odyssey* was written in a meter called *dactylic hexameter,* named for the Latin or Greek words for finger. The words link the

three main bones in the finger (one long and two short) to the form of one long and two short syllables repeated six times (hexameter, *hex* meaning "six"). In fact, often the last part of a line is long-short. To understand what dactylic hexameter sounds like in *The Odyssey,* you can either read the Greek (you first) or turn to Rodney Merrill's English translation, which maintains the dactylic hexameter (now, that is truly heroic): Merrill R. The Rhythm of the Epic. http://www.home.earthlink.net/~merrill_odyssey/id5.html. Accessed December 28, 2018.

Perhaps not surprisingly, given the limitation of the meter, Homer would use a few tricks to keep the story flowing. One was to use multiple different epithets for his main characters. So Odysseus was variously referred to as "Man of Pain" or "Man of Action." Sometimes, he was "Raider of Cities." Other times, he was "Loved of Zeus."

101 *entered the common lexicon* Use of the word *odyssey* in the English literature corpus from 1800 is available through: Google Books Ngram Viewer. https://books.google.com/ngrams.

## 9. The Luck of the Irish

103 *Matt Might had a blog* The Might family stories come from many conversations with Matt Might, including one referenced specifically in the text on November 14, 2018. References were also drawn from Seth Mnookin's excellent article in *The New Yorker* and other more recent articles: Mnookin S. One of a Kind. *New Yorker.* 2014;21:32-38; Weintraub K. A Battle Plan for a War on Rare Diseases. *New York Times.* https://www.nytimes.com/2018/09/10/health/matthew-might-rare-diseases.html. Published September 10, 2018. Accessed May 3, 2020; The Might of the Mights: Parents Overcome Genetics to Save Son—Rare Genomics Institute. Rare Genomics Institute. https://www.raregenomics.org/blog/2016/4/10/the-might-of-the-mights-parents-overcome-genetics-to-save-son. Published April 10, 2016. Accessed May 3, 2020.

104 *he started two tech companies* Matt Might started two companies. One grew out of his thesis work on automated reasoning to protect security vulnerabilities, and another was a smaller project, theapplet.com. This was a chat room app that took advantage of the emerging embedding of JavaScript in web pages to provide interactivity. It ended up being used by *Newsweek, The Washington Post,* and others to host interactive chat rooms.

104 *teamed up on a pioneering study* The original paper from Duke including the first report of *NGLY1* as a potential disease entity: Need AC, Shashi V, Hitomi Y, et al. Clinical application of exome sequencing in undiagnosed genetic conditions. *J Med Genet.* 2012;49:353-361.

106 *"lack of tears" or "dry eyes"* A lack of tears, or alacrima, is in fact a recognized feature of only one other syndrome, the so-called triple A syndrome comprising achalasia (a disorder of the esophagus or food pipe), Addison's disease (a disorder of production of the hormone cortisol), and alacrima. It was clear early on that Bertrand's diagnosis was not triple A syndrome.

106 *you need Liam Neeson* Liam Neeson starred in the 2008 movie directed by Pierre Morel, *Taken:* Besson L. *Taken.* Los Angeles: 20th Century Fox; 2009.

Matt Might's blog: Might M. Hunting down my son's killer. http://matt.might.net/articles/my-sons-killer/. Retrieved July 2015.

107 *Matt and Kristen Wilsey* Details came from personal conversations with Matt and Kristen Wilsey, including one on December 22, 2018, and follow-up emails and conversations. Details also from this article in: Hawk S. With Grace. Stanford Graduate School of Business. https://www.gsb.stanford.edu/insights/grace. Published October 18, 2019. Accessed May 3, 2020.

109 *John Freidenrich* John Freidenrich was a former chairman of the board of trustees of Stanford and played a major role in Stanford hospitals. As well as giving his name to a building in which some of our research is now carried out, he once gave me one of the best pieces of advice I ever received for Mother's Day (ditch the fancy restaurant for a picnic in the park!). Long-time Stanford Leader, Donor John Freidenrich Dies. *Stanford News.* https://news.stanford

.edu/2017/10/18/leader-donor-john-freidenrich-dies/. Published October 18, 2017. Accessed May 3, 2020.

110 *Matthew Bainbridge* Details of the work from Baylor, including that of Matthew Bainbridge, come from a conversation with Matthew Bainbridge on February 3, 2020, and follow-up emails.

111 *Bainbridge immediately emailed Kristen Wilsey* Matthew Bainbridge and Matt Wilsey kindly shared emails from this time, including the one where they described Grace's clinical presentation (February 26, 2013).

112 *including a family from Turkey* As well as Matthew Bainbridge, the blog was found by Yale geneticist Murat Gunel, who was sequencing patients with undiagnosed disease from Turkey. His mind immediately went to two siblings whose cases his team had been trying to solve. They had similar features to those described in the blog post, including developmental delay and movement disorder. He had sequenced those two siblings, so he pulled up the file with the list of possible variants in candidate genes that had arisen from his analysis of their genomes and found *NGLY1*. Caglayan AO, Comu S, Baranoski JF, et al. NGLY1 mutation causes neuromotor impairment, intellectual disability, and neuropathy. *Eur J Med Genet.* 2015;58(1):39-43.

## 10. Next-Day Delivery

113 *diagnosing a disease* The paper describing the disease resulting from mutations in *NGLY1:* Enns GM, Shashi V, Bainbridge M, et al. Mutations in NGLY1 cause an inherited disorder of the endoplasmic reticulum-associated degradation pathway. *Genet Med.* 2014;16(10):1-8.

And the accompanying editorial written by Matt Might and Matt Wilsey: Might M, Wilsey M. The shifting model in clinical diagnostics: How next-generation sequencing and families are altering the way rare diseases are discovered, studied, and treated. *Genet Med.* 2014;16(10):1-2.

116 *Chow made a fruit fly model with the fly* NGLY1 *gene disrupted* This is described in this paper: Owings KG, Lowry JB, Bi Y, Might M, Chow CY. Transcriptome and functional analysis in a Drosophila model of NGLY1 deficiency provides insight into therapeutic approaches. *Hum Mol Genet.* 2018;27(6):1055-1066.

117 *in vitro fertilization* The Edwards and Steptoe paper in *Nature* that reported in vitro fertilization: Edwards RG, Steptoe PC, Purdy JM. Fertilization and cleavage in vitro of preovulator human oocytes. *Nature.* 1970;227:1307.

117 *Louise Brown* Louise Brown was teased at school for being a "test tube" baby. She later went on to have her own children.

117 *sperm in a lab dish* A little-known fact is that, left to their own devices, sperm swim in circles: Friedrich BM, Jülicher F. The stochastic dance of circling sperm cells: Sperm chemotaxis in the plane. *New J Phys.* 2008;10(12):123025; Kaupp UB. 100 years of sperm chemotaxis. *J Cell Biol.* 2012;199(6):i9-i9; Friedrich BM, Jülicher F. Chemotaxis of sperm cells. *Proc Natl Acad Sci USA.* 2007;104(33):13256-13261.

117 *preimplantation genetic diagnosis* A nice summary is found here: Handyside AH. Preimplantation genetic diagnosis after 20 years. *Reprod Biomed Online.* 2010;21(3):280-282.

118 *Grace Science Foundation* Grace Science Foundation. https://gracescience.org/. Accessed May 3, 2020. The Might family also set up a foundation (NGLY1.org).

118 *to study NRF1* Tomlin FM, Gerling-Driessen UIM, Liu Y-C, et al. Inhibition of NGLY1 inactivates the transcription factor Nrf1 and potentiates proteasome inhibitor cytotoxicity. *ACS Cent Sci.* 2017;3(11):1143-1155.

119 *the scientist who originally discovered the* NGLY1 *gene* Suzuki T, Kwofie MA, Lennarz WJ. Ngly1, a mouse gene encoding a deglycosylating enzyme implicated in proteasomal degradation: Expression, genomic organization, and chromosomal mapping. *Biochem Biophys Res Commun.* 2003;304(2):326-332.

119 *the expression of ENGase* Huang C, Harada Y, Hosomi A, et al. Endo-β-N-acetylglucosaminidase forms N-GlcNAc protein aggregates during ER-associated degradation in Ngly1-defective cells. *Proc Natl Acad Sci USA.* 2015;112(5):1398-1403.

119 *a team at the University of Utah* Bi Y, Might M, Vankayalapati H, Kuberan B. Repurposing of proton pump inhibitors as first identified small molecule inhibitors of endo-β-N-acetylglucosaminidase (ENGase) for the treatment of NGLY1 deficiency, a rare genetic disease. *Bioorg Med Chem Lett.* 2017;27(13):2962-2966.

119 *Recent work from Hudson Freeze* Tambe MA, Ng BG, Freeze HH. N-Glycanase 1 transcriptionally regulates aquaporins independent of its enzymatic activity. *Cell Rep.* 2019;29(13):4620-4631.e4.

## 11. Hoofbeats in Central Park

121 *we published a paper* When we launched the Undiagnosed Diseases Network, we wrote about the plans: Gahl WA, Wise AL, Ashley EA. The Undiagnosed Diseases Network of the National Institutes of Health: A national extension. *JAMA.* 2015;314(17):1797-1798; Ramoni RB, Mulvihill JJ, Adams DR, et al. The Undiagnosed Diseases Network: Accelerating discovery about health and disease. *Am J Hum Genet.* 2017;100(2):185-192.

After the first 1,519 patients, we wrote up the findings: Splinter K, Adams DR, Bacino CA, et al. Effect of genetic diagnosis on patients with previously undiagnosed disease. *N Engl J Med.* 2018;379(22):2131-2139.

122 *We first met Danny and Nikki Miller* Details come from our many interactions with the family over the course of 2017 to 2019 and beyond. In addition, I undertook a recorded interview with Danny by teleconference on December 20, 2018.

125 *One key paper* Heimer G, Kerätär JM, Riley LG, et al. MECR mutations cause childhood-onset dystonia and optic atrophy, a mitochondrial fatty acid synthesis disorder. *Am J Hum Genet.* 2016;99(6):1229-1244.

127 *They appeared on* CBS This Morning *and on NPR's* Morning Edition Medical Detectives: The Last Hope for Families Coping with Rare Diseases. NPR. https://www.npr.org/sections/health-shots/2018/12/17/673066806/medical-detectives-the-last-hope-for-families-coping-with-rare-diseases. Published December 17, 2018. Accessed May 16, 2020; "Doctor Detectives" Help Diagnose Mysterious Illnesses with DNA Analysis. CBS News. https://www.cbsnews.com/news/undiagnosed-diseases-network-dna-helps-miller-family-diagnose-mepan-syndrome/. Published October 11, 2018. Accessed May 16, 2020.

127 *the MEPAN foundation* MEPAN Founcation. https://www.mepan.org/.

128 *Matchmaker Exchange* Matchmaker Exchange. https://www.matchmakerexchange.org/.

130 *Schnitzler syndrome* Auto-inflammatory conditions, including Schnitzler syndrome: Palladini G, Merlini G. The elusive pathogenesis of Schnitzler syndrome. *Blood.* 2018;131(9):944-946.

133 *Occam's razor* William of Ockham's actual words appear to be only loosely connected to the commonly interpreted meaning: "Nothing ought to be posited without a reason given, unless it is self-evident or known by experience or proved by the authority of Sacred Scripture." Spade PV, Panaccio C. William of Ockham. In Zalta EN, ed. *The Stanford Encyclopedia of Philosophy.* Winter 2016. Stanford, CA: Stanford University; 2016. https://plato.stanford.edu/archives/win2016/entries/ockham/.

134 *Central Park* Interestingly, while Central Park Zoo doesn't have zebras, it does have snow leopards.

134 *horses, not zebras* When You Hear Hoofbeats Look for Horses Not Zebras. Quote Investigator. https://quoteinvestigator.com/2017/11/26/zebras/. Accessed December 14, 2017.

134 *To put this in perspective* The chance of being struck by lightning is about one in three thousand: Flash Facts About Lightning. *National Geographic.* June 2005. https://news.nationalgeographic.com/news/2004/06/flash-facts-about-lightning/. Accessed January 4, 2019.

The odds of winning the $1.6 billion lottery Powerball were around one in three hundred million: Yan H. We're Not Saying You Shouldn't Play, But Here Are 5 Things More Likely To Happen Than You Winning the Lottery. CNN. October 2018. https://www.cnn.com/2018/10/23/us/lottery-winning-odds-trnd/index.html. Accessed January 4, 2019.

The Stanford blog post about the one-in-a-billion patient: Digitale E, Ford A, Hite E. Stanford Team Helps Patient Who Is "Unique in the World." Scope. https://scopeblog.stanford.edu/2016/12/14/stanford-team-helps-patient-who-is-unique-in-the-world/. Published December 14, 2016. Accessed January 4, 2019.

The paper reporting the findings: Zastrow DB, Zornio PA, Dries A, et al. Exome sequencing identifies de novo pathogenic variants in FBN1 and TRPS1 in a patient with a complex connective tissue phenotype. *Cold Spring Harb Mol Case Stud.* 2017;3(1):a001388.

138 *reported the discovery of a new disease* The new mitochondrial syndrome is caused by the gene *ATP5F1D.* The paper is here: Oláhová M, Yoon WH, Thompson K, et al. Biallelic mutations in ATP5F1D, which encodes a subunit of ATP synthase, cause a metabolic disorder. *Am J Hum Genet.* 2018;102(3):494-504.

The syndrome definition is here: OMIM Entry-# 618120 Mitochondrial complex V (ATP synthase) deficiency, nuclear type 5; MC5DN5 OMIM. https://www.omim.org/entry/618120. Accessed January 8, 2019.

Anahi was featured in the *San Francisco Chronicle* story: Allday E. "Disease Detectives" Crack Cases of 130 Patients with Mysterious Illnesses. *San Francisco Chronicle.* https://www.sfchronicle.com/health/article/Disease-detectives-crack-cases-of-130-13297547.php. Published October 11, 2018. Accessed January 8, 2019.

139 *featured some of our patients in an article* Kolata G. When the Illness Is a Mystery, Patients Turn to These Detectives. *New York Times.* https://www.nytimes.com/2019/01/07/health/patients-medical-mysteries.html. Published January 7, 2019. Accessed May 16, 2020.

## PART III: AFFAIRS OF THE HEART

### 12. Whisky à Go-Go

145 *when German obstetrician Friedrich Ludwig Meissner* This story comes from Meissner's book, that I acquired in the original German. Meissner FL. *Taubstummheit, Ohr-u. gehörkrankheiten: Bd. 1. Taubstummheit u. Taubstummenbildung.* C. F. Winter'sche Verlagshandlung, Leipzig & Heidelberg Winter; 1856.

The section relating to the little girl's sudden death was translated kindly for me by my neighbor Jenny Suckale, a professor of geophysics at Stanford.

145 *Anton Jervell and Fred Lange-Nielsen* The history of Jervell and Lange-Nielsen: Jervell A, Lange-Nielsen F. Congenital deaf-mutism, functional heart disease with prolongation of the QT interval, and sudden death. *Am Heart J.* 1957. http://www.sciencedirect.com/science/article/pii/0002870357900790; Tranebjaerg L, Bathen J, Tyson J, Bitner-Glindzicz M. Jervell and Lange-Nielsen syndrome: A Norwegian perspective. *Am J Med Genet.* 1999;89(3):137-146.

Fred Lange-Nielsen played bass in the SwingKlubb Band in Oslo in the 1940s: Evensmo J. The Altosaxes of Swing in Norway. http://www.jazzarcheology.com/artists/swing_in_norway.pdf. Updated October 6, 2011.

146 *Owen Conor Ward* Details of the contributions of Owen Conor Ward and Caesaro Romano come from a retrospective in the *Irish Medical Journal:* Hodkinson EC, Hill AP, Vandenberg JI. The Romano-Ward syndrome—1964-2014: 50 years of progress. *Ir Med J.* 2014;107(4):122-124.

Details of the eight-year-old girl are derived from an interview with Owen Conor Ward published in the *Irish Times:* Hunter N. A Medical Stalwart Now in Happy Exile. http://www.irishhealth.com/article.html?id=18437. Accessed April 4, 2017.

Interestingly, Conor Ward moved to London later in his career and became very interested in medical history. He spent four years writing a biography of John Langdon Down (who gave his name to Down syndrome). *John Langdon Down, A Caring Pioneer* was published by the Royal Society of Medicine.

146 *torsades de pointes* In the most commonly accepted version, defined by Dessertenne in 1966: Dessertenne F, et al. La tachycardie ventriculaire a deux foyers opposes variables. *Arch Mal Coeur Vaiss.* 1966;59(2):263-272. Both *torsades* and *pointes* are plural, thus the correct form of the connecting word should likely be plural *des:* Mullins ME. Mon bête noir (my pet peeve). *J Med Toxicol.* 2011;7(2):181. However, Dessertenne himself used a singular *de* at times, so it likely this one will be fought about by rhythm and language scholars for a while.

149 *we even contacted a local drug company* We contacted Gilead.

150 *Stephen Kingsmore* Kingsmore's rapid turnaround protocols for genome sequencing in the neonatal intensive care unit are described here: Saunders CJ, Miller NA, Soden SE, et al. Rapid whole-genome sequencing for genetic disease diagnosis in neonatal intensive care units. *Sci Transl Med.* 2012;4(154):154ra135-ra154ra135.

151 *Kyla Dunn* Kyla Dunn worked on pieces for PBS's *Frontline, NOVA ScienceNOW,* and CBS's *60 Minutes II* and had articles in, among other publications, *The Atlantic, The Washington Post, The New York Times Book Review,* and *Discover* magazine. She appears in IMDb as a producer on several science documentaries, including *Frontline, 60 Minutes, NOVA ScienceNOW,* and *Wired Science:* Kyla Dunn. IMDb. https://www.imdb.com/name/nm1871408/. Accessed January 9, 2019.

Her *Nature* paper is here: Gibbs CS, Coutré SE, Tsiang M, et al. Conversion of thrombin into an anticoagulant by protein engineering. *Nature.* 1995;378(6555):413-416.

Unlike Kyla, I do not appear on IMDb, but I did record with my saxophone quartet a version of Charles Mingus's "Haitian Fight Song" for the Ewan MacGregor movie *Young Adam* (music director: David Byrne). Sadly, there was a technical problem with the version we recorded, and by the time of the redo, I had moved away to do residency at Oxford. My other forlorn hope for movie immortality was related to the Pixar movie *Brave.* Pixar hired a Stanford linguistics professor to do high-definition facial video motion capture so they could capture the movements associated with a West of Scotland accent for the lead character (a feisty red-haired heroine named Merida). Our clinic manager stayed till the very last credit of the movie, but to her dismay, my name was not to be found.

153 *a way that resembled the go-go dancing* The human ether à go-go story comes from a *Science News* article that quotes William Kaplan directly: Weiss R. Mutant monikers. *Sci News.* 1991;139(2):30-31.

The origin of go-go dancing is unclear, as is the role of the Whisky à Go-Go bar in originating the phenomenon. The French term *à go go* means " a lot" or "in abundance." Note the bar uses the Scottish spelling of *whisky* without the *e,* whereas several of the articles focused on the bar use the spelling more commonly associated with Irish whiskey or bourbon.

154 *more "precise" medicine* Work on precision therapy for long QT 3 has been led by Sylvia Priori, Michael Ackerman, and others: Priori SG, Wilde AA, Horie M, et al. HRS/EHRA/APHRS expert consensus statement on the diagnosis and management of patients with inherited primary arrhythmia syndromes: Document endorsed by HRS, EHRA, and APHRS in May 2013 and by ACCF, AHA, PACES, and AEPC in June 2013. *Heart Rhythm.* 2013;10(12):1932-1963; Schwartz PJ, Priori SG, Locati EH, et al. Long QT syndrome patients with mutations of the SCN5A and HERG genes have differential responses to Na+ channel blockade and to increases in heart rate. Implications for gene-specific therapy. *Circula-*

*tion.* 1995;92(12):3381-3386; Mazzanti A, Maragna R, Faragli A, et al. Gene-specific therapy with mexiletine reduces arrhythmic events in patients with long QT syndrome type 3. *J Am Coll Cardiol.* 2016;67(9):1053-1058.

## 13. How Many Genomes Are You?

156 *then a whole bunch come at once* The science of bus bunching is fairly well characterized. Here's a good example describing a statistical approach, including mixture modeling: Ma Z, Ferreira L, Mesbah M, Zhu S. Modeling distributions of travel time variability for bus operations. *J Adv Transp.* 2016;50(1):6-24.

And another exploring potential solutions: Verbich D, Diab E, El-Geneidy A. Have they bunched yet? An exploratory study of the impacts of bus bunching on dwell and running times. *Public Transp.* 2016;8(2):225-242.

156 *Although none of us can explain it* Cognitive theories of medical learning and diagnostics are discussed in: Schmidt HG, Norman GR, Boshuizen HP. A cognitive perspective on medical expertise: theory and implication [published erratum appears in *Acad Med* 1992 Apr;67(4):287]. *Acad Med.* 1990;65(10):611-621.

And a commentary I wrote as a medical student: Ashley EA. Medical education—beyond tomorrow? The new doctor—Asclepiad or Logiatros? *Med Educ.* 2000;34(6):455-459.

158 *five heads out of twenty-five flips by chance alone* The probability of seeing five reads out of twenty-five is calculated according to the density function of the binomial distribution (readily calculated using the statistical program R and the function *dbinom*).

159 *had not been observed for a genetic cardiovascular disease before* Although not relating to a genetic cardiac disease, cardiac mosaicism of the connexin 40 gene had been suggested by a paper in *The New England Journal of Medicine* for the common condition atrial fibrillation where it was found in three individuals after sequencing tissue fixed in formalin and paraffin embedded. Thibodeau IL, Xu J, Li Q, et al. Paradigm of genetic mosaicism and lone atrial fibrillation: Physiological characterization of a connexin 43-deletion mutant identified from atrial tissue. *Circulation.* 2010;122(3):236-244; Gollob MH, Jones DL, Krahn AD, et al. Somatic mutations in the connexin 40 gene (GJA5) in atrial fibrillation. *N Engl J Med.* 2006;354(25):2677-2688.

Other investigators unfortunately failed to find it: Roberts JD, Longoria J, Poon A, et al. Targeted deep sequencing reveals no definitive evidence for somatic mosaicism in atrial fibrillation. *Circ Cardiovasc Genet.* 2015;8(1):50-57. Although as our experience shows, the phenomenon is rare.

Some have speculated that the variants might have arisen as an artifact of sequencing tissue fixed in formalin: Chen L, Liu P, Evans TC, Ettwiller LM. DNA damage is a major cause of sequencing errors, directly confounding variant identification. *bioRxiv.* 2016. http://www.biorxiv.org/content/early/2016/08/19/070334.abstract.

160 somatic mosaicism Somatic mosaicism has also been recognized associated with the special process of cell division required to make sperm and eggs. This leads to something called *gonadal* mosaicism (the testes and ovaries are the *gonads*) and is one way that a child can appear to have a "new" genetic variant not present in or at least not apparent from the blood of the parents. In this case, testing the sperm or eggs of the parents would reveal the mutation.

161 *neurofibromatosis* An excellent general overview of mosaicism including a discussion of neurofibromatosis type 1 is found in: Biesecker LG, Spinner NB. A genomic view of mosaicism and human disease. *Nat Rev Genet.* 2013;14(5):307-320.

Other nice summaries on somatic mosaicism: Poduri A, Evrony GD, Cai X, Walsh CA. Somatic mutation, genomic variation, and neurological disease. *Science.* 2013;341(6141):1237758; Lupski JR. Genetics. Genome mosaicism—one human, multiple genomes. *Science.*

2013;341(6144):358-359; Forsberg LA, Gisselsson D, Dumanski JP. Mosaicism in health and disease—clones picking up speed. *Nat Rev Genet.* 2017;18(2):128-142.

161 *Your bone marrow* Clonal hematopoiesis as a risk factor for heart attack: Jaiswal S, Fontanillas P, Flannick J, et al. Age-related clonal hematopoiesis associated with adverse outcomes. *N Engl J Med.* 2014;371(26):2488-2498; Fuster JJ, Walsh K. Somatic mutations and clonal hematopoiesis: Unexpected potential new drivers of age-related cardiovascular disease. *Circ Res.* 2018;122(3):523-532; Jaiswal S, Natarajan P, Silver AJ, et al. Clonal hematopoiesis and risk of atherosclerotic cardiovascular disease. *N Engl J Med.* 2017;377(2):111-121.

162 *such as one reported by science journalist Carl Zimmer* Carl Zimmer's piece on mosaicism and chimerism: Zimmer C. DNA Double Take. *New York Times.* https://www.nytimes.com/2013/09/17/science/dna-double-take.html. Published September 16, 2013. Accessed December 15, 2017.

Also featuring work of Mike Snyder and Alex Urban: O'Huallachain M, Karczewski KJ, Weissman SM, Urban AE, Snyder MP. Extensive genetic variation in somatic human tissues. *Proc Natl Acad Sci USA.* 2012;109(44):18018-18023.

Of note, Astrea's story is featured in Carl Zimmer's excellent book: Zimmer C. *She Has Her Mother's Laugh: The Powers, Perversions, and Potential of Heredity.* New York: Penguin; 2018.

162 *who published his own whole genome* James Lupski's paper sequencing his own genome: Lupski JR, Reid JG, Gonzaga-Jauregui C, et al. Whole-genome sequencing in a patient with Charcot-Marie-Tooth neuropathy. *N Engl J Med.* 2010;362(13):1181-1191.

Lupskis's writing on mosaicism: Lupski JR. Genetics. Genome mosaicism—one human, multiple genomes. *Science.* 2013;341(6144):358-359.

166 *Natalia Trayanova* Biographical detail on Natalia Trayanova comes from emails with Natalia and this online interview: Natalia Trayanova on Developing Computer Simulations of Hearts. Johns Hopkins Medicine. https://www.hopkinsmedicine.org/research/advancements-in-research/fundamentals/profiles/natalia-trayanova. Accessed December 15, 2017.

Modeling of biological systems is summarized nicely by Brian Ingalls: Ingalls BP. *Mathematical Modeling in Systems Biology: An Introduction.* Cambridge, MA: MIT Press; 2013.

## 14. Shake, Rattle, and Roll

172 *the largest ever removed was over three hundred pounds* In fact, the largest cystic tumor ever removed was 303 pounds. It was removed by Professor Katherine O'Hanlan of Stanford in 1991. The surgery is actually on YouTube (of course it is). Kate O'Hanlan, MD. World's Largest Tumor. https://www.youtube.com/watch?v=wwiN_TbpqMA. Published May 27, 2012. Accessed May 17, 2020.

173 *Carney complex* The earliest description of Carney complex was in: Young WF, Carney JA, Musa BU, Wulffraat NM, Lens JW, Drexhage HA. Familial Cushing's syndrome due to primary pigmented nodular adrenocortical disease. *N Engl J Med.* 1989;321(24):1659-1664.

174 *J. Aidan Carney* Carney's biography is from this short book chapter: Dy BM, Lee GS, Richards ML. J. Aidan Carney. In Pasieka JL, Lee JA, eds. *Surgical Endocrinopathies.* Boston: Springer; 2015:229-231. Aidan Carney told Constantine Stratakis (referenced later) that he came from Mayo County, Ireland, not Roscommon County. He thought this was amazing because he ended up working at the Mayo Clinic (obviously, the Mayo family was of Irish descent). The Wikipedia article about him states that he was born in Roscommon County. The two counties are neighboring, so it is possible that he was born in one and grew up in the other. Carney's personal odyssey is described in his own words here: Carney JA. Discovery of the Carney complex, a familial lentiginosis–multiple endocrine neoplasia syndrome: A medical odyssey. *Endocrinologist.* 2003;13(1):23.

177 *was one of the most influential papers in genetics* Botstein's paper and a review a decade

later: Botstein D. Using the genetic linkage map of the human genome to understand complex inherited diseases. *J Nerv Ment Dis.* 1989;177(10):644. doi:10.1097/00005053-198910000-00012; Botstein D, White RL, Skolnick M, Davis RW. Construction of a genetic linkage map in man using restriction fragment length polymorphisms. *Am J Hum Genet.* 1980;32(3):314-331.

178 *Stratakis and Carney reported* Many of these details come from a personal conversation with Constantine Stratakis. I spoke with him by telephone on January 8, 2019.

The original paper and others: Correa R, Salpea P, Stratakis CA. Carney complex: An update. *Eur J Endocrinol.* 2015;173(4):M85-M97; Stratakis CA, Kirschner LS, Carney JA. Clinical and molecular features of the Carney complex: Diagnostic criteria and recommendations for patient evaluation. *J Clin Endocrinol Metab.* 2001;86(9):4041-4046; Kirschner LS, Carney JA, Pack SD, et al. Mutations of the gene encoding the protein kinase A type I-α regulatory subunit in patients with the Carney complex. *Nat Genet.* 2000;26(1).

180 *gigantism* One of the larger international studies of gigantism: Rostomyan L, Daly AF, Petrossians P, et al. Clinical and genetic characterization of pituitary gigantism: An international collaborative study in 208 patients. *Endocr Relat Cancer.* 2015;22(5):745-757.

180 *Transsphenoidal resection of the pituitary* During long days at the endocrine unit at the Radcliffe hospital in Oxford during my residency, my fellow resident (Suzannah Wilson, a talented future cardiologist) and I used to amuse ourselves by making up little songs. We were amazed by these pituitary surgeries, performed on patients referred by renowned endocrinologist John Wass, by a surgeon named C. B. T. Adams. Since only his initials appeared on the surgery lists, we knew him as CBTA, which, of course, we pronounced as "ciabatta." The lyrics, if I recall, went something like: *Ciabatta, Ciabatta / what a wonderful surgeon is he / John Wass, John Wass / he's the one for acromegaly.* To make it fit the meter, we had to lengthen the *a* in *Wass* and the *e* in *acromegaly,* so it wasn't our best work, but on pretty minimal sleep, you do what you can.

184 *I had never heard of someone donating their heart to a relative* There have in fact been cases reported where a heart was donated from one relative to another. Chester Szuber was a Christmas tree farmer who had suffered multiple heart attacks and had been on the transplant list for four years. Patti, his daughter, was training to be a nurse and was on vacation when she was involved in a serious car accident. Her parents were called with the bad news that she was about to pass and were asked to go to the hospital to certify her identity and sign the paperwork for her to donate organs. Patti was a strong advocate for organ donation and a proud carrier of a donor card. In the midst of their grief, Chester's family asked the obvious question. Could her heart be donated to her father? Chester's immediate reaction was, "No way," but after a lot of thought and particularly, after his whole family convinced him that this is exactly what his daughter would have wanted, he agreed. Patti's heart has lived twenty-two more years in her dad's chest. Patti's brother, Bob, commented that he believed giving her heart to her dad would make his sister "the happiest little angel in heaven." Schaefer J. A Few Minutes with . . . a Father with a Full Heart. *Detroit Free Press.* https://www.freep.com/story/news/columnists/jim-schaefer/2016/07/16/chester-szuber-heart-transplant/87098726/. Published July 17, 2016. Accessed December 13, 2017; Father Receives Heart Transplant from Daughter. *New York Times.* http://www.nytimes.com/1994/08/26/us/father-receives-heart-transplant-from-daughter.html. Published August 26, 1994. Accessed December 13, 2017.

The Organ Procurement and Transplantation network clarified the legality of directed organ donation here: OPTN Information Regarding Deceased Directed Donation. OPTN. https://optn.transplant.hrsa.gov/news/optn-information-regarding-deceased-directed-donation/. Accessed December 14, 2017.

Questions had arisen around Steve Jobs's liver transplant after several people, including Tim Cook, had offered to donate a partial liver directly to him: Gayomali C. Apple CEO Tim Cook Tried to Give Steve Jobs His Liver—But Jobs Refused. *Fast Company.* https://www

.fastcompany.com/3043628/apple-ceo-tim-cook-tried-to-give-steve-jobs-his-liver-but-jobs-refused. Published March 12, 2015. Accessed December 14, 2017.

I had one patient ask me once if he could donate his heart to his wife, who was dying of heart failure. He was very serious, and when I clarified that he would be giving his life to save hers, he confirmed that was what he was proposing. Fifty years of marriage and devotion, he could see he was losing the love of his life to heart failure. I melted inside, turned away, and worked very hard to hold my composure as I explained that actively taking one life to save another was not something we could do.

184 *who had also recently died* Ricky's cousins' deaths were not related to the heart, so not connected to his genetic condition.

186 *Jonas Korlach* Pacific Biosciences history comes from conversations with Jonas Korlach and Steve Turner, as well as from this article: MacKenzie RJ. A SMRTer Way to Sequence DNA? Genomics Research from Technology Networks. https://www.technologynetworks.com/genomics/articles/a-smrter-way-to-sequence-dna-309952. Published September 25, 2018. Accessed May 17, 2020.

188 *The answer was right there in front of us* Our paper describing the first long-read, whole-genome sequencing in a patient is here: Merker JD, Wenger AM, Sneddon T, et al. Long-read genome sequencing identifies causal structural variation in a Mendelian disease. *Genet Med.* June 2017. doi:10.1038/gim.2017.86.

## 15. River of the Land of Pine Trees

196 *Susan Graham's daughter Leilani* I have had the privilege to follow Leilani's extraordinary story since she was seventeen. In addition, a recorded interview with Susan and Chris Graham on March 24, 2017, and recorded interviews with Leilani on December 19, 2016, and January 21, 2017, are sources of historical details.

197 *generation growing up increasingly sedentary* Childhood obesity statistics from the CDC: Hales CM, Carroll MD, Fryar CD, Ogden CL. Prevalence of Obesity Among Adults and Youth: United States, 2015–2016. *NCHS Data Brief.* 2017;(288):1-8.

197 *the earliest insights into hypertrophic cardiomyopathy* It was either Laennec or René-Joseph Hyacinth Bertin who first coined the term *hypertrophy.* Duffin J. *To See with a Better Eye: A Life of R. T. H. Laennec.* Princeton, NJ: Princeton University Press; 2014.

Laennec devoted a lot of his early work to the anatomy of the heart. He substituted the words *hypertrophy* and *dilatation* for his teacher Jean-Nicolas Corvisart's use of *active* and *passive* aneurysms of the heart. Why he switched is not clear. Laennec discussed with Bertin which of them had the priority on the use of the word *hypertrophy,* the latter claiming he coined it in a communication in August 1811. Laennec countered that he had used the word *hypertrophy* for some time but also conceded that he might not have invented it. In the end, Laennec called it a draw, writing to his friend, "It is quite natural that in a matter of pure and simple observation, two men carefully examine the same object and see the same."

198 *In probably the first description of hypertrophic cardiomyopathy* Historical references are found in: McKenna WJ, Sen-Chowdhry S. From Teare to the present day: A fifty year odyssey in hypertrophic cardiomyopathy, a paradigm for the logic of the discovery process. *Rev Esp Cardiol.* 2008;61(12):1239-1244; Adelman H, Adelman A. The logic of discovery a case study of hypertrophic cardiomyopathy. *Acta Biotheor.* 1977;26(1):39-58; Mirchandani S, Phoon CKL. Sudden cardiac death: A 2400-year-old diagnosis? *Int J Cardiol.* 2003;90(1):41-48. Also, Coats and Hollman Heart (http://dx.doi.org/10.1136/hrt.2008.153452)

Anatomic pathology arose in the 1500s during the time of Andreas Vesalius. The first description of hypertrophic cardiomyopathy was from Giovanni Battista Morgagni, a pathologist working in the early 1700s who quoted Théophile Bonet, a Swiss physician from Geneva describing a coachman who had died suddenly in his carriage (Théophile Bonet, *Sepulchretum seu anatomia practica* [Geneva, 1679]). French pathologists Henri Liouville and Henri Hallopeau described a series of cases, including a seventy-five-year-old woman with symptoms of heart failure and an unusual cardiac murmur with a heart wall that was four times the normal width.

198 *it was another pathologist, Donald Teare* Background on Teare is from: Watkins H, Ashrafian H, McKenna WJ. The genetics of hypertrophic cardiomyopathy: Teare redux. *Heart.* 2008;94(10):1264-1268. In his seminal paper, Teare connected "living" and "postmortem" findings via the electrocardiogram. ECG strips viewed on those who later died, revealed findings in seven out of eight patients who died suddenly (one from a stroke, likely related to a non-life-threatening rhythm called *atrial fibrillation*). The family mentioned in Teare's first and seminal paper was so intriguing, they became the focus of a more detailed paper, published in 1960. Goodwin JF, Hollman A, Cleland WP, Teare D. Obstructive cardiomyopathy simulating aortic stenosis. *Br Heart J.* 1960;22:403-414.

199 *Wood was pale, acerbic, and short, with steely blue eyes* Paul Wood's biography: Silverman ME, Somerville W. To die in one's prime: The story of Paul Wood. *Am J Cardiol.* 2000;85(1):75-88; Camm J. The contributions of Paul Wood to clinical cardiology. *Heart Lung Circ.* 2003;12 Suppl 1:S10-S14.

Paul Wood founded the field of invasive cardiology as well as, probably, cardiac surgery in Britain, and he contributed enormously to our understanding of many cardiac diseases. Gaston Bauer said of him, "He took in pupils and he turned out disciples." In describing the physical examination in hypertrophic cardiomyopathy, he also described some maneuvers that could be performed to bring out these findings more clearly, such as squatting then standing.

199 *the outflow tract obstruction* The dynamic, short-lived obstruction to blood flow leaving the heart appeared in a case described by William Harvey. The mitral valve (*mitral* rhyming with *title*—it is named for its similarity in shape to a Christian bishop's ceremonial headgear, the miter). In addition to Russell Brock, another group from Washington University School of Medicine in Saint Louis noted the similarity of this dynamic obstruction to aortic obstruction referring to it as *pseudo* aortic stenosis.

199 *had taken place in the UK* Myectomies were performed in the UK by William Cleland and Russell Brock (Goodwin JF, Hollman A, Cleland WP, Teare D. Obstructive cardiomyopathy simulating aortic stenosis. Br Heart J 1960;22:403–14; Russell Brock: Brock R, Fleming PR. Aortic subvalvar stenosis; a report of 5 cases diagnosed during life. *Guys Hosp Rep.* 1956;105[4]:391-408) and the United States (Andrew Glenn Morrow) at similar times. Braunwald E. Hypertrophic cardiomyopathy: The first century 1869-1969. *Glob Cardiol Sci Pract.* 2012;2012(1):5.

The story of the US surgery I heard directly from Eugene Braunwald himself (personal interview, October 2, 2018) and at a Hypertrophic Cardiomyopathy Association meeting in New Jersey. The story is also told in print: Maron BJ, Braunwald E. Eugene Braunwald, MD and the early years of hypertrophic cardiomyopathy: A conversation with Dr. Barry J. Maron. *Am J Cardiol.* March 2012. doi:10.1016/j.amjcard.2012.01.376; Maron BJ, Roberts WC. The father of septal myectomy for obstructive HCM, who also had HCM: The unbelievable story. *J Am Coll Cardiol.* 2016;67(24):2900-2903.

200 *Morrow was later himself diagnosed with hypertrophic cardiomyopathy* These details and quotations come from a personal interview with Eugene Braunwald on October 2, 2018, at Stanford University. Similar details are found in this article: Maron BJ, Braunwald E. Eugene Braunwald, MD and the early years of hypertrophic cardiomyopathy: A conversation with Dr. Barry J. Maron. *Am J Cardiol.* March 2012. doi:10.1016/j.amjcard.2012.01.376.

201 *Barry Maron, from the United States, and William "Bill" McKenna from the UK* Barry Maron and Bill McKenna have published over one thousand papers between them. The guideline document and this *New England Journal of Medicine* review paper are two of the smaller number they published together: Spirito P, Seidman CE, McKenna WJ, Maron BJ. The management

of hypertrophic cardiomyopathy. *N Engl J Med.* 1997;336(11):775-785; Maron BJ, McKenna WJ, Danielson GK, et al. American College of Cardiology / European Society of Cardiology Clinical Expert Consensus Document on Hypertrophic Cardiomyopathy. A report of the American College of Cardiology Foundation Task Force on Clinical Expert Consensus Documents and the European Society of Cardiology Committee for Practice Guidelines. *Eur Heart J.* 2003;24(21):1965-1991.

201 *systolic anterior motion* Cardiologist Douglas Wigle in Canada is usually credited with discovering the forward movement of the mitral valve causing obstruction in his original references to *muscular subaortic stenosis.* Stanford cardiologist Richard "Rich" Popp formalized the echocardiographic criteria in 1974.

202 *which helps to form new memories* Details on short-term memory: Gluck MA, Mercado E, Myers CE. *Learning and Memory: From Brain to Behavior*. New York: Macmillan Higher Education; 2007.

202 *hydrogen peroxide* This is the same chemical used to dye hair blond (also it is an antiseptic).

204 *It begins with Peter Christian Abildgaard* The earliest reports of collapse ("swooning") and sudden death come from Hippocrates in 400 B.C.: Hippocrates, Coar T. *The Aphorisms of Hippocrates: with a Translation into Latin and English.* Omaha, NE: Classics of Medicine Library; 1982.

Abildgaard's work presented in 1775 was translated from the original Latin and published in Thomas Driscol's paper in 1975: Driscol TE, Ratnoff OD, Nygaard OF. The remarkable Dr. Abildgaard and countershock. The bicentennial of his electrical experiments on animals. *Ann Intern Med.* 1975;83(6):878-882; Abildgaard PC. Tentamina electrica in animalibus instituta. *Societatis Medicae Havniensis Collectanea.* 1775;2:157.

205 *Benjamin Franklin* Franklin's famous kite experiment involved drawing electricity from storm clouds into a Leyden jar. In a parallel to Abildgaard's work, he wrote in a letter to Peter Collinson in 1748 about using electric shocks to kill a turkey: "A turkey is to be killed for our dinner by the electrical shock, and roasted by the electrical jack, before a fire kindled by the electrified bottle." Franklin is often credited with the first use of the word *battery* in an electrical context. "As six jars, however, discharged at once, are capable of giving a very violent shock, the operator must be very circumspect, lest he should happen to make the experiment on his own flesh, instead of that of the fowl" (*Memoirs of Benjamin Franklin,* volume 2, page 328, letter to Messrs Dubourg and d'Alibard concerning the mode of rendering meat tender by electricity). In another letter, Franklin intriguingly even refers to resurrection of a fowl or turkey cock "as promised by M Dubourg." It is however unclear what Dr. Dubourg had in mind as a mechanism for this resurrection.

"I wish it were possible, from this instance to invent a method of embalming drowned persons in such a manner that they may be recalled to life at any period however distant; for having a very ardent desire to see and observe the state of America an hundred years hence, I should prefer to any ordinary death, the being immersed in a cask of Madeira wine, with a few friends till that time, to be then recalled to life by the solar warmth of my dear country! But since in all probability we live in an age too early and too near the infancy of science, to hope to see such an art brought in our time to its perfection, I must for the present content myself with the treat, which you are so kind as to promise me, of the resurrection of a fowl or a turkey-cock" (Memoirs of Benjamin Franklin, volume 2, pages 391-2, letter to M Dubourg).

205 *So naturally, Abildgaard built his own* Some have disputed whether the Leyden jars would have contained sufficient charge to shock the heart of a chicken to ventricular fibrillation via a shock applied to the head. Also, it is known that smaller hearts do not maintain ventricular fibrillation well, leading some to suggest that this was neurogenic fibrillation and spontaneous cardioversion. Certainly it is clear why the shock would not have been enough to down a horse. Akselrod H, Kroll MW, Orlov MV. History of Defibrillation. In Efimov IR, Kroll MW, Tchou PJ, eds. *Cardiac Bioelectric Therapy.* Boston: Springer; 2009:15-40.

206 *Sophie Greenhill* The story of the young girl is told on the Royal College of Physicians website

and here: Akselrod H, Kroll MW, Orlov MV. History of Defibrillation. In Efimov IR, Kroll MW, Tchou PJ, eds. *Cardiac Bioelectric Therapy.* Boston: Springer; 2009:15-40. These authors question whether a fall could have caused ventricular fibrillation.

207 *James Curry* There is a copy of James Curry's book *Observations on Apparent Death from Drowning, Suffocation Etc* in the Royal College of Physicians, published in London in 1790: The First Defibrillator? The Work of James Curry. RCP London. https://www.rcplondon.ac.uk/news/first-defibrillator-work-james-curry. Published May 19, 2017. Accessed January 14, 2018.

207 *Charles Kite* More on James Curry and Charles Kite is found in: Hurt R. Modern cardiopulmonary resuscitation—not so new after all. *J R Soc Med.* 2005;98(7):327-331; Cakulev I, Efimov IR, Waldo AL. Cardioversion: Past, present, and future. *Circulation.* 2009;120(16):1623-1632.

Kite's book is available in the University of Leeds archive: An Essay on the Recovery of the Apparently Dead: Kite, Charles, 1768–1811. Internet Archive. https://archive.org/details/b21510829. Accessed January 14, 2018.

208 *Do Not Resuscitate tattoo* Do not resuscitate tattoo: Holt GE, Sarmento B, Kett D, Goodman KW. An unconscious patient with a DNR tattoo. *N Engl J Med.* 2017;377(22):2192-2193.

208 *the first medically documented internal defibrillation* The first medically documented internal defibrillation is described here: Beck CS, Pritchard WH, Feil HS. Ventricular fibrillation of long duration abolished by electric shock. *J Am Med Assoc.* 1947;135(15):985.

209 *push hard and fast* A good aide-mémoire comes from this American Heart Association video featuring the physician-comedian Ken Jeong and using the Bee Gees' disco hit "Stayin' Alive": https://www.youtube.com/watch?v=iXcsHoQMGqc.

Ken Jeong's medical training and comedy is described in this *Washington Post* article: Horton A. A Woman Had A Seizure at Ken Jeong's Comedy Show. The Former Doctor Jumped Offstage to Save Her. *Washington Post.* https://www.washingtonpost.com/news/arts-and-entertainment/wp/2018/05/07/a-woman-had-a-seizure-at-ken-jeongs-comedy-show-the-former-doctor-jumped-offstage-to-save-her/. Published May 7, 2018. Accessed May 7, 2018.

210 *a casino* Casinos, defibrillators, and sudden death: Valenzuela TD, Roe DJ, Nichol G, Clark LL, Spaite DW, Hardman RG. Outcomes of rapid defibrillation by security officers after cardiac arrest in casinos. *N Engl J Med.* 2000;343(17):1206-1209.

211 *Mordechai Frydman* The story of the first implantable ICD: Mirowski M, Reid PR, Mower MM, et al. Termination of malignant ventricular arrhythmias with an implanted automatic defibrillator in human beings. *N Engl J Med.* 1980;303(6):322-324.

213 *captured the imagination of the medical community* The sight of Mirowski's dog was so miraculous to some that they even questioned its veracity, asking how long it took to train the dog to collapse like that.

213 *Anthony Van Loo* Anthony Van Loo's cardiac arrest is on YouTube, and the link here is to a video annotated by one of my patients: Hugo Campos. Soccer Player Anthony Van Loo Survives a Sudden Cardiac Arrest (SCA) When His ICD Fires. (ANNOTATED). https://www.youtube.com/watch?v=DU_i0ZzIV5U. Published June 12, 2009. Accessed December 7, 2017.Another CPR and automatic external defibrillator shock in a young volleyball player is shown here: https://www.youtube.com/watch?v=MtHZ6ItHiTc

217 *understanding the cause of disease in this family* Peter Paré's paper on the Coaticook family: Paré JAP, Fraser RG, Pirozynski WJ, Shanks JA, Stubington D. Hereditary cardiovascular dysplasia: A form of familial cardiomyopathy. *Am J Med.* 1961;31(1):37-62.

218 *the discovery was seminal* The classic discovery papers on the genetic basis of hypertrophic cardiomyopathy: Jarcho JA, McKenna W, Pare JA, et al. Mapping a gene for familial hypertrophic cardiomyopathy to chromosome 14q1. *N Engl J Med.* 1989;321(20):1372-1378; Tanigawa G, Jarcho JA, Kass S, et al. A molecular basis for familial

hypertrophic cardiomyopathy: An alpha/beta cardiac myosin heavy chain hybrid gene. *Cell.* 1990;62(5):991-998; Geisterfer-Lowrance A a., Kass S, Tanigawa G, et al. A molecular basis for familial hypertrophic cardiomyopathy: A beta cardiac myosin heavy chain gene missense mutation. *Cell.* 1990;62(5):999-1006.

218 *The following years led to the discovery* Stories from these early days in the elucidation of the genetic basis of HCM come from Hugh Watkins, Calum MacRae and others in the Seidman lab. Follow-up papers on the genetic basis of HCM.: Watkins H, Rosenzweig A, Hwang DS, et al. Characteristics and prognostic implications of myosin missense mutations in familial hypertrophic cardiomyopathy. *N Engl J Med.* 1992;326(17):1108-1114; Watkins H, McKenna WJ, Thierfelder L, et al. Mutations in the genes for cardiac troponin T and alpha-tropomyosin in hypertrophic cardiomyopathy. *N Engl J Med.* 1995;332(16):1058-1064; Niimura H, Bachinski LL, Sangwatanaroj S, et al. Mutations in the gene for cardiac myosin-binding protein C and late-onset familial hypertrophic cardiomyopathy. *N Engl J Med.* 1998;338(18):1248-1257.

## 16. Songs in the Key of Life

229 *with mortality rates of 10–20 percent* Intensive care unit mortality: Capuzzo M, Volta C, Tassinati T, et al. Hospital mortality of adults admitted to Intensive Care Units in hospitals with and without Intermediate Care Units: A multicentre European cohort study. *Crit Care.* 2014;18(5):551.

231 *exercise, and metabolic factors* Leslie Leinwand's running mice with hypertrophic cardiomyopathy demonstrated that treadmill running could actually reverse some of the fibrosis and disarray. Konhilas JP, Watson PA, Maass A, et al. Exercise can prevent and reverse the severity of hypertrophic cardiomyopathy. *Circ Res.* 2006;98(4):540-548.

No other intervention at the time had been shown capable of reversing any aspect of the disease. We actually set out, together with colleagues at the University of Michigan, to test this idea in humans and performed the first randomized study of exercise training in HCM. Published in 2017 in the *Journal of the American Medical Association* and led by our Michigan colleagues Sharlene Day and Sara Saberi, our study showed that not only was exercise safe for HCM patients but that it could increase the output from the heart after just a short period of exercise training. Saberi S, Wheeler M, Bragg-Gresham J, et al. Effect of moderate-intensity exercise training on peak oxygen consumption in patients with hypertrophic cardiomyopathy: A randomized clinical trial. *JAMA.* 2017;317(13):1349-1357.

231 *cardiac growth in well-fed pythons* Leslie Leinwand's python work was inspired by Jared Diamond: Secor SM, Diamond J. Effects of meal size on postprandial responses in juvenile Burmese pythons (Python molurus). *Am J Physiol.* 1997;272(3 Pt 2):R902-R912; Andersen JB, Rourke BC, Caiozzo VJ, Bennett AF, Hicks JW. Physiology: Postprandial cardiac hypertrophy in pythons. *Nature.* 2005;434(7029):37-38; Riquelme C a., Magida J a., Harrison BC, et al. Fatty acids identified in the Burmese python promote beneficial cardiac growth. *Science.* 2011;334(6055):528-531.

235 *an idea that came to a collaborator in a dream* Jim Spudich telling his myosin story: iBiology. James Spudich (Stanford) 4: Myosin mutations and hypertrophic cardiomyopathy. https://www.youtube.com/watch?v=-zqUUo_qmTM. Posted November 1, 2017.

The myosin mesa told in scientific papers and narratives: Nag S, Trivedi DV, Sarkar SS, et al. The myosin mesa and the basis of hypercontractility caused by hypertrophic cardiomyopathy mutations. *Nat Struct Mol Biol.* 2017;24(6):525-533; Spudich JA. The myosin mesa and a possible unifying hypothesis for the molecular basis of human hypertrophic cardiomyopathy. *Biochem Soc Trans.* 2015;43:64-72; Trivedi DV, Adhikari AS, Sarkar SS, Ruppel KM, Spudich JA. Hypertrophic cardiomyopathy and the myosin mesa: Viewing an old disease in a new light. *Biophys Rev.* 2018;10(1):27-48.

Modeling of genetic variation in patients and the population underlining the importance of the myosin mesa: Homburger JR, Green EM, Caleshu C, et al. Multidimensional structure-function relationships in human β-cardiac myosin from population-scale genetic variation. *Proc Natl Acad Sci USA.* 2016;113(24):6701-6706.

241 *parallels with Eva Cassidy* The hauntingly beautiful "Over the Rainbow" from Eva Cassidy: https://www.youtube.com/watch?v=2rd8VktT8xY

## PART IV: PRECISELY ACCURATE MEDICINE

### 17. Superhumans

245 *Mäntyranta started winning cross-country ski events at an early age* Papers on the Mäntyranta family: La DE, de la Chapelle A, Träskelin a. L, Juvonen E. Truncated erythropoietin receptor causes dominantly inherited benign human erythrocytosis. *Proc Natl Acad Sci USA.* 1993;90(10):4495-4499; de la Chapelle A, Sistonen P, Lehväslaiho H, Ikkala E, Juvonen E. Familial erythrocytosis genetically linked to erythropoietin receptor gene. *Lancet.* 1993;341(8837):82-84; Juvonen E, Ikkala E, Fyhrquist F, Ruutu T. Autosomal dominant erythrocytosis caused by increased sensitivity to erythropoietin. *Blood.* 1991;78(11):3066-3069.

247 *David Epstein, author of the book* The Sports Gene David Epstein's fantastic book: Epstein DJ. *The Sports Gene: Inside the Science of Extraordinary Athletic Performance.* New York: Penguin; 2014.

249 Exercise at the Limit—Inherited Traits of Endurance Maria Konnikova: Superhero Genes: What Sets the World's Most Elite Athletes Apart? *California Sunday Magazine.* https://story.californiasunday.com/superhero-gene-euan-ashley-stanford. Published August 4, 2016. Accessed August 25, 2019.

ELITE study feature: Wilner J. Can Superhuman Athletes Provide Genetic Clues on Heart Health? *Mercury News.* https://www.mercurynews.com/2017/10/29/4851089/. Published October 29, 2017. Accessed August 25, 2019.

250 *Norwegian cross-country skier* There are a couple of reasons beyond the possibility of favorable genetics that help explain why so many Scandinavians are among the fittest people in the world, at least as judged by VO2max. One is that the highest values tend to come from those sports where the largest muscle mass is recruited. Although elite runners and cyclists tend to exhibit very high VO2max, these athletes focus on keeping their upper bodies "quiet" while their lower bodies power their forward motion, but in sports like cross-country skiing, the whole-body muscle mass is recruited. Another reason may be that exercise testing and athlete physiology had its foundations in Scandinavia, in particular in the Åstrand Laboratory of Work Physiology at the Karolinska Institute, so many more Scandinavian athletes were tested in the early days. This was in fact where Mikael Mattsson completed his Ph.D.

251 *the same gene that helps Andean highlanders* Chronic mountain sickness in Peru: Gazal S, Espinoza JR, Austerlitz F, et al. The genetic architecture of chronic mountain sickness in Peru. *Front Genet.* 2019;10:690.

251 *critical for the ability of their cells to produce energy* NAD metabolism: Yaku K, Okabe K, Nakagawa T. NAD metabolism: Implications in aging and longevity. *Ageing Res Rev.* 2018;47:1-17.

251 *Sharlayne Tracy Nature* editorial describing Sharlayne Tracy (note this is not her real name): Hall SS. Genetics: A gene of rare effect. *Nature.* 2013;496(7444):152-155.

252 *French geneticist Catherine Boileau* Catherine Boileau's discovery of *PCSK9* as a cause of familial hypercholesterolemia: Abifadel M, Varret M, Rabès J-P, et al. Mutations in *PCSK9* cause autosomal dominant hypercholesterolemia. *Nature Genetics.* 2003;34(2):154-156. doi:10.1038/

ng1161; Varret M, Rabès J-P, Saint-Jore B, et al. A third major locus for autosomal dominant hypercholesterolemia maps to 1p34.1-p32. *Am J Hum Genet.* 1999;64(5):1378-1387.

Some details also come from personal conversations and emails with Catherine Boileau.

253 *Helen Hobbs and Jonathan Cohen* Cohen J, Pertsemlidis A, Kotowski IK, Graham R, Garcia CK, Hobbs HH. Low LDL cholesterol in individuals of African descent resulting from frequent nonsense mutations in PCSK9. *Nat Genet.* 2005;37(2):161-165; Zhao Z, Tuakli-Wosornu Y, Lagace TA, et al. Molecular characterization of loss-of-function mutations in PCSK9 and identification of a compound heterozygote. *Am J Hum Genet.* 2006;79(3):514-523; Cohen JC, Boerwinkle E, Mosley TH Jr, Hobbs HH. Sequence variations in PCSK9, low LDL, and protection against coronary heart disease. *N Engl J Med.* 2006;354(12):1264-1272.

Details also come from personal emails and this interview with Helen Hobbs: A Conversation with Helen Hobbs. *Journal of Clinical Investigation.* https://www.jci.org/articles/view/84086. Published October 1, 2015.

255 *famed drug developer and Merck CEO-to-be P. Roy Vagelos* Robert Plenge alerted me to the story from the 1970s from P. Roy Vagelos. Vagelos's son Randall (Randy) Vagelos is a cardiologist, friend, and mentor at Stanford who taught me everything I know about heart failure and critical care cardiology.

256 *Robert Plenge wrote in the journal* Nature Reviews Drug Discovery The influential paper: Plenge RM, Scolnick EM, Altshuler D. Validating therapeutic targets through human genetics. *Nat Rev Drug Discov.* 2013;12(8):581-594.

260 *Sek Kathiresan* Sek Kathiresan's seminal mendelian randomization paper: Voight BF, Peloso GM, Orho-Melander M, et al. Plasma HDL cholesterol and risk of myocardial infarction: A mendelian randomisation study. *Lancet.* 2012;380(9841):572-580.

261 *the BBC correspondent David Cox* The BBC feature on pain: Cox D. The curse of the people who never feel pain. BBC. http://www.bbc.com/future/story/20170426-the-people-who-never-feel-any-pain. Accessed August 26, 2019.

263 *venomous spiders* Drugs derived from tarantula toxin: Xu H, Li T, Rohou A, et al. Structural basis of Nav1.7 inhibition by a gating-modifier spider toxin. *Cell.* 2019;176(4):702-715.e14.

## 18. Precision Medicine

264 *Eric Dishman* Eric Dishman's story came from various meetings with him at Intel, at the Stanford campus, and in Washington, D.C. I completed a recorded interview with him on September 10, 2019. Some information is drawn from his TED Talk: Dishman E. *Health Care Should Be a Team Sport.* https://www.ted.com/talks/eric_dishman_health_care_should_be_a_team_sport?language=en. Accessed June 7, 2020.

Stanford talk: Stanford Medicine. Eric Dishman, NIH-Stanford Medicine Big Data | Precision Health 2018. https://www.youtube.com/watch?v=P4qjP4VVp_c. Published June 21, 2018. Accessed June 7, 2020.

267 *Andy Grove* Details of Andy Grove's life come from conversations with Eric Dishman and Sean Maloney. And Tedlow RS. *Andy Grove: The Life and Times of an American Business Icon.* New York: Penguin; 2007; Also articles: Rivett-Carnac M. The True Story of Andrew Grove, Time's 1997 Man of the Year. *Time.* March 2016. https://time.com/4267150/andrew-grove-intel-survivor-biography-budapest/. Accessed June 7, 2020; Andrew Grove: A Survivor's Tale. *Time.* http://content.time.com/time/magazine/article/0,9171,987588,00.html. Accessed June 7, 2020; Kandell J. Andrew S. Grove Dies at 79; Intel Chief Spurred Semiconductor Revolution. *New York Times.* https://www.nytimes.com/2016/03/22/technology/andrew-grove-intel-obituary.html. Published March 22, 2016. Accessed June 7, 2020.

269 *gathered together critical thinkers* Others involved in the Collins group included: Joan Bailey-Wilson, NHGRI; Greg Burke, Wake Forest; Chris Hook, Mayo Clinic; Rod Howell, NICHD; Jean MacCluer, Southwest Foundation; Don Mattison, NICHD; Jeff Murray, University

of Iowa; Larry Needham, CDC; Anne Spence, UC–Irvine; Alec Wilson, NHGRI; Sam Wilson, NIEHS.

271 *United Kingdom launched its own ambitious effort* Details of this history of the UK Biobank come from personal conversations and emails with John Bell, Rory Collins, and Mark McCarthy. John Bell's paper appears here: Bell J. The new genetics in clinical practice. *BMJ.* 1998;316(7131):618-620.

George Poste article: Fears R, Poste G. Policy forum: Health care delivery. Building populations genetics resources using the U.K. NHS. *Science.* 1999;284(5412):267-268.

272 *wrote up an opinion piece for the journal* Nature Collins FS. The case for a US prospective cohort study of genes and environment. *Nature.* 2004;429(6990):475-477.

273 *Barack Obama, the junior senator from Illinois* Details come from recorded interviews with Jennifer Leib (September 12, 2019), Dora Hughes (September 29, 2019), and a conversation with Eduardo Ramos (December 19, 2019).

273 *with the exception of a run-in* This story came from Barack Obama's book *Dreams From My Father:* Obama B. *Dreams from My Father: A Story of Race and Inheritance.* Edinburgh, UK: Canongate Books; 2007.

274 *the bill never saw the light of day* Obama's original bill did not pass, but by all accounts, Obama's leadership on the issue was helpful in increasing interest and awareness of the possibilities/potential for genomics to transform medical care. Even after his departure from the Senate, modified versions of the bill continued to be introduced, and it influenced others (notably Ted Kennedy) to introduce related legislation.

275 *Jill Holdren* Details of Jill's story come from emails with her and her dad, John Holdren, and a personal conversation on May 21, 2020. Jill also mentioned an encounter with *BRCA* pioneer Mary-Claire King. John Holdren was unavailable for a dinner at the National Academy of Sciences, and his wife suggested his daughter accompany her instead. Jill was introduced almost immediately to Mary-Claire King, and after a few moments, found herself sharing her personal story with the physician-scientist. The two started talking and didn't stop for almost two hours, during which time, the breast and ovarian cancer pioneer offered help and advice as well as to be her treating physician if she wanted to move to Seattle. For those who know her, this is typical of her passion and personal investment. A great insight into her life is provided by this account: Who Can You Trust? https://themoth.org/stories/who-can-you-trust.

276 *Jill suggested Obama should himself be tested* Some details of Obama's mother's ovarian cancer come from this video from his sister Maya Soetoro-Ng. ovariancancerorg. Dr. Maya Soetoro-Ng Ovarian Cancer PSA, Full Version. https://www.youtube.com/watch?v=EQmM7QQyvgs. Published September 15, 2015. Accessed June 7, 2020. She mentions in the video that she and her brother were both tested and found to be negative for *BRCA* mutations.

277 *Obama shook hands on both sides of the aisle* Details come from the official video: Obama White House. President Obama's 2015 State of the Union Address. https://www.youtube.com/watch?v=cse5cCGuHmE. Published January 20, 2015. Accessed June 7, 2020.

Official transcript: Obama B. Remarks by the President in State of the Union Address January 20, 2015. White House. https://obamawhitehouse.archives.gov/the-press-office/2015/01/20/remarks-president-state-union-address-January-20-2015.

278 *ad lib* State of the Union remarks as prepared: President Obama's State of the Union Address—Remarks As Prepared for Delivery. Medium. https://medium.com/@ObamaWhiteHouse/president-obamas-state-of-the-union-address-remarks-as-prepared-for-delivery-55f9825449b2. Published January 21, 2015. Accessed September 9, 2019.

State of the Union 2015, remarks as delivered: https://www.youtube.com/watch?v=cse5cCGuHmE&t=16s. Retrieved. 2011;27:2011; Remarks by the President in State of the Union Address | January 20, 2015. White House. https://obamawhitehouse.archives.gov/the-press-office/2015/01/20/remarks-president-state-union-address-January-20-2015. Published January 20, 2015. Accessed September 9, 2019.

278 *the official era of precision medicine had begun* The term *precision medicine* first came to prominence in 2011 in a publication from the National Research Council. National Research Council (US) Committee on a Framework for Developing a New Taxonomy of Disease. *Toward Precision Medicine: Building a Knowledge Network for Biomedical Research and a New Taxonomy of Disease.* Washington, D.C.: National Academies Press; 2012; Ashley EA. Towards precision medicine. *Nat Rev Genet.* 2016;17:507.

An expert panel was convened to focus on building a knowledge network for biomedical research and a new taxonomy of disease (a *taxonomy* is a classification system). The report was titled, *Toward Precision Medicine.* One of the members of the writing group was the former chair of pathology at Stanford, Stephen Galli, who explained to me that one of his contributions was to write the appendix entry clarifying the committee's use of the term *precision medicine* as opposed to the more traditional *personalized medicine.* Personalized medicine was sometimes narrowly interpreted as meaning that individual treatments were being designed for individuals. A different term, it was felt, would help emphasize the group's focus on a new classification of disease.

278 *In the East Room that day* Press event for the launch of the Precision Medicine Initiative: Remarks by the President on Precision Medicine. White House. https://obamawhitehouse.archives.gov/the-press-office/2015/01/30/remarks-president-precision-medicine. Published January 30, 2015. Accessed September 9, 2019.

Details come from Francis Collins, Eric Green, Zak Kohane, and others. The story of the DNA model comes from a personal conversation with Eric Green.

279 *a great example of precision medicine in action* The relationship between the cystic fibrosis foundation and Vertex Pharmaceuticals is described here: Tozzi J. This Medical Charity Made $3.3 Billion from a Single Pill. *Bloomberg News.* https://www.bloomberg.com/news/features/2015-07-07/this-medical-charity-made-3-3-billion-from-a-single-pill. Published July 7, 2015. Accessed September 9, 2019.

Ivacaftor was developed by Vertex. The example of cystic fibrosis is particularly relevant for Francis Collins, because in 1989, he and collaborator Lap-Chee Tsui reported the cause of cystic fibrosis, the first major genetic disease to have its genetic cause elucidated. The gene in question, *CFTR,* is a chloride channel: it pumps charged molecules of chloride in and out of specific cells, particularly in the lungs and digestive tract. When it fails to work properly, sticky secretions cause problems with lung infections and the digestion of food. After the gene was identified and patients started being routinely sequenced for variants in *CFTR,* it became apparent that patients could be subdivided into recognizable groups. In some patients, the mutant channel would travel appropriately to the cell surface but, when it got there, would not function properly. In other patients, the mutant channel never reached the cell surface, because it was tagged as abnormal and recycled by the cell. The new drug increased the probability that the chloride channels on the cell surface were open, something that would obviously work a lot better in the subgroup where the mutant channel actually made it to the cell surface. So that was the initial group in whom the drug was tested. This kind of targeting is the essence of precision medicine.

281 *it became known as All of Us* All of Us Research Program Investigators, Denny JC, Rutter JL, et al. The "All of Us" Research Program. *N Engl J Med.* 2019;381(7):668-676.

282 *Kári Stefánsson* Master Decoder: A Profile of Kári Stefánsson. *Scientist.* https://www.the-scientist.com/profile/master-decoder—a-profile-of-kri-stefnsson-65517. Accessed September 9, 2019.

282 *an enormous online database* Iceland dating app: Buckley C. There's an App That Keeps Icelanders from Dating Their Relatives. Culture Trip. https://theculturetrip.com/europe/iceland/articles/iceland-is-so-small-theres-an-app-that-keeps-icelanders-from-dating-their-relatives/. Accessed September 16, 2019.

The name is actually a reference to the original Íslendingabók, a written record of the county's earliest history—including the genealogy of its founders. The author was an Icelandic priest, Ari Þorgilsson, working in the early twelfth century.

283 *Erik Ingelsson* Erik Ingelsson is a physician and geneticist who was at Stanford when this work was published in *The Lancet:* Ganna A, Ingelsson E. 5 year mortality predictors in 498 103 UK Biobank participants: A prospective population-based study. *Lancet.* 2015;386(9993):533-540.

283 *an online engine* Try it out: Global Biobank Engine. https://biobankengine.stanford.edu/.

284 *the British proclivity toward tea, coffee, and alcohol* Coffee, tea, and alcohol and bitter taste perception: Ong J-S, Hwang DL-D, Zhong VW, et al. Understanding the role of bitter taste perception in coffee, tea and alcohol consumption through Mendelian randomization. *Sci Rep.* 2018;8(1):16414.

284 *Not someone prone to compliments or hyperbole* Stefánsson on UK biobank: https://twitter.com/anderson_carl/status/1176142417864605696?s=20. September 23, 2019, International Common Disease Alliance inaugural meeting.

285 *of which 23andMe* Details come from personal conversations with Anne Wojcicki, Richard Scheller, and Robert Gentleman (the cocreator of the influential statistical programming language R).

287 *Daniel MacArthur* Details come from personal conversations with Dan, with postdoc Konrad Karczewski (who completed his Ph.D. in Mike Snyder's lab), and from the GnomAD website. Francioli L, Tiao G, Karczewski K, Solomonson M, Watts N. gnomAD v2. 1. MacArthur Lab. 2018. https://gnomad.broadinstitute.org/.

## 19. Genome Surgery

292 *Holbrook Kohrt* Details of this remarkable young man come from personal interactions with him and conversations with his contemporaries and mentors. Email permission to tell his story came from his mother, Marylou Kidd (Februrary 14, 2020). Also some published articles: Roberts S. Dr. Holbrook Kohrt, Hemophiliac Who Made Condition a Crusade, Dies at 38. *New York Times.* https://www.nytimes.com/2016/03/02/health/dr-holbrook-kohrt-hemophiliac-who-made-the-condition-a-crusade-dies-at-38.html. Published March 1, 2016. Accessed June 7, 2020; Snyder A. Holbrook Kohrt. *Lancet.* 2016;387(10030):1810.

294 *Jesse Gelsinger* Details are from: The Death of Jesse Gelsinger, 20 Years Later. Science History Institute. https://www.sciencehistory.org/distillations/the-death-of-jesse-gelsinger-20-years-later. Published June 4, 2019. Accessed June 7, 2020; Gene-therapy trials must proceed with caution. *Nature.* 2016;534(7609):590.

294 The Washington Post *reported in 1999* Weiss R, Nelson D. Teen dies undergoing experimental gene therapy. *Washington Post.* 1999; https://www.washingtonpost.com/wp-srv/WPcap/1999-09/29/060r-092999-idx.html. Accessed August 9, 2020.

296 *clinical trials in hemophilia to reach completion* Sources on gene therapy for hemophilia: Dunbar CE, High KA, Joung JK, Kohn DB, Ozawa K, Sadelain M. Gene therapy comes of age. *Science.* 2018;359(6372). doi:10.1126/science.aan4672; Rangarajan S, Walsh L, Lester W, et al. AAV5-factor VIII gene transfer in severe hemophilia A. *N Engl J Med.* 2017;377(26):2519-2530; VandenDriessche T, Chuah MK. Hyperactive factor IX Padua: A game-changer for hemophilia gene therapy. *Mol Ther.* 2018;26(1):14-16; Pasi KJ, Fischer K, Ragni M, et al. Long-term safety and sustained efficacy for up to 5 years of treatment with recombinant factor IX Fc fusion protein in subjects with haemophilia B: Results from the B-YOND extension study. *Haemophilia.* June 2020. doi:10.1111/hae.14036

297 *Consider the Guardino family* Details including quotes come from online sources: America's Got Talent 2017 Christian Guardino Just the Intro and Judges' Comments S12E03; 2017. https://www.youtube.com/watch?v=pSjXKpGdXBw. Accessed January 2, 2020; Howard C. "AGT" Contestant Born with Blinding Disease Says Gift of Sight Allows Him "to See Such Incredible Things." August 2019. Fox News. https://www.foxnews.com/health/agt-contestant-born-blinding-disease-sight. Accessed January 12, 2020.

298 *The results were dramatic* The eye was an early target for gene therapy because of its accessibility and because of something called *ocular immune privilege:* Streilein JW. Ocular immune privilege: Therapeutic opportunities from an experiment of nature. *Nat Rev Immunol.* 2003;3(11):879-889; Taylor AW. Ocular immune privilege. *Eye.* 2009;23(10):1885-1889.

This refers to the phenomenon whereby genetic therapy delivered by virus is less likely to be inactivated because the body avoids triggering severe immune reactions in the eye.

299 *Spinal muscular atrophy* Spinal muscular atrophy: Mercuri E, Darras BT, Chiriboga CA, et al. Nusinersen versus sham control in later-onset spinal muscular atrophy. *N Engl J Med.* 2018;378(7):625-635; Mendell JR, Al-Zaidy S, Shell R, et al. Single-dose gene-replacement therapy for spinal muscular atrophy. *N Engl J Med.* 2017;377(18):1713-1722; Lorson CL, Hahnen E, Androphy EJ, Wirth B. A single nucleotide in the SMN gene regulates splicing and is responsible for spinal muscular atrophy. *Proc Natl Acad Sci USA.* 1999;96(11):6307-6311; Kashima T, Manley JL. A negative element in SMN2 exon 7 inhibits splicing in spinal muscular atrophy. *Nat Genet.* 2003;34(4):460-463.

304 *My own interest in RNA silencing began in 2006* Our work in this area is the focus of presentations and publications: Zaleta K, Wheeler MT, Finsterbach T, Ashley EA. Allele specific silencing in vivo in a model of hypertrophic restrictive cardiomyopathy. Presented at *Keystone Meeting: Cardiovascular Genetics*. Tahoe City, California; 2013; Zaleta-Rivera K, Dainis A, Ribeiro AJS, et al. Allele-specific silencing ameliorates restrictive cardiomyopathy attributable to a human myosin regulatory light chain mutation. *Circulation.* 2019;140(9):765-778; Zaleta K, Wheeler M, Finsterbach TP, Ashley EA. Allele specific silencing of mutant alleles in hypertrophic cardiomyopathy. *J RNAi Gene Silencing.* 2013;9:486-489.

The story about my earliest conversation with Andy Fire was also told by Hanae Armitage at Stanford Communications: Armitage H, Dusheck J, Goldman B, Huber J, Stankus K. "Turning Down the Volume" of a Faulty Gene in Heart Disease. Scope. https://scopeblog.stanford.edu/2019/08/20/turning-down-the-volume-of-a-faulty-gene-in-heart-disease/. Published August 20, 2019. Accessed June 7, 2020.

308 *otherwise known as severe combined immunodeficiency* Why SCID gene therapy led to leukemia: Why Gene Therapy Caused Leukemia in Some "Boy in the Bubble Syndrome" Patients. *Science Daily.* August 2008. https://www.sciencedaily.com/releases/2008/08/080807175438.htm. Accessed February 13, 2020.

311 *The CRISPR story has its origins in the 1990s* Historical detail from original sources, Jennifer Doudna's book: Doudna JA, Sternberg SH. *A Crack in Creation: Gene Editing and the Unthinkable Power to Control Evolution.* Boston: Houghton Mifflin Harcourt; 2017; and Eric Lander's article Lander ES. The Heroes of CRISPR. *Cell.* 2016;164(1-2):18-28.

As well as personal conversations with Jennifer Doudna, Feng Zhang, and David Liu.

313 *With the science of gene editing advancing* While we are waiting for major therapeutic advances in human disease using CRISPR, one major application has been in muscular dystrophy, where a CRISPR system called *Cpf1* has been used by Eric Olson, a renowned cardiovascular biologist, to correct mutations in a mouse model. In a paper published in 2017, he showed that after injecting the gene correction cocktail into young embryos, the mice later showed significantly improved muscle strength and reduced muscle damage compared with their untreated siblings.

314 *Francis Collins, the head of the NIH* Francis Collins's statement on CRISPR babies: Collins FS. Statement on Claim of First Gene-Edited Babies by Chinese Researcher. National Institutes of Health. 2018. https://www.nih.gov/about-nih/who-we-are/nih-director/statements/statement-claim-first-gene-edited-babies-chinese-researcher.

315 *Mila Makovec loved the outdoors* Details from personal conversations with Tim Yu and interactions with Mila's mom, Julia, Tim Yu, and our Stanford team. Also from published sources: Keshavan M, Branswell H, Herper M, Joseph A. Saving Mila: How Doctors Raced to Stop a Young Girl's Rare Disease. STAT. https://www.statnews.com/2018/10/22/a-tailor-made-therapy-may-have-halted-a-rare-disease/. Published October 22, 2018. Accessed June 7,

2020; Kim J, Hu C, Moufawad El Achkar C, et al. Patient-customized oligonucleotide therapy for a rare genetic disease. *N Engl J Med.* 2019;381(17):1644-1652.

## 20. The Road Ahead

321 *food insecurity* Sequencing to protect food security: Boykin LM, Ghalab A, De Marchi BR, et al. Real time portable genome sequencing for global food security. *bioRxiv.* May 2018:314526. doi:10.1101/314526.

Nanopore sequencing of foodborne pathogens: Taylor TL, Volkening JD, DeJesus E, et al. Rapid, multiplexed, whole genome and plasmid sequencing of foodborne pathogens using long-read nanopore technology. *Sci Rep.* 2019;9(1):16350.

322 *Excitingly, that day has finally come* We wrote about this in more detail here: Knowles JW, Ashley EA. Cardiovascular disease: The rise of the genetic risk score. *PLOS Med.* 2018;15(3):e1002546.

326 *forty trillion bacteria* You can't do better than Ed Yong's amazing book *I Contain Multitudes* if you want to read more about our synergistic relationship with microorganisms: Yong E. *I Contain Multitudes: The Microbes Within Us and a Grander View of Life.* New York: Random House; 2016. Yong was also the single most compelling scientific correspondent during the pandemic. See, for example: https://www.theatlantic.com/magazine/archive/2020/09/coronavirus-american-failure/614191/

326 *spilled over* The superb David Quammen book *Spillover* describes this phenomenon in detail: Quammen D. *Spillover: Animal Infections and the Next Human Pandemic.* New York: W. W. Norton & Company; 2012.

326 *Google searches* Data from Google Trends: Coronavirus Expontential Growth. Google Trends. https://trends.google.com/trends/explore?geo=US&q=coronavirus%20exponential%20growth. Accessed June 9, 2020.

327 *lock down* Data from: Glanz J, Robertson C. Lockdown Delays Cost at Least 36,000 Lives, Data Show. *New York Times.* https://www.nytimes.com/2020/05/20/us/coronavirus-distancing-deaths.html. Published May 21, 2020. Accessed June 9, 2020; Pei S, Kandula S, Shaman J. Differential effects of intervention timing on COVID-19 spread in the United States. *medRxiv.* May 2020. doi:10.1101/2020.05.15.20103655.

327 *1918 flu pandemic* Details from: Spinney L. *Pale Rider: The Spanish Flu of 1918 and How It Changed the World.* New York: PublicAffairs; 2017.

327 *their digital ink was even dry.* The pandemic also saw the rise of the "preprint" in biomedical science. This refers to a version of a scientific article uploaded by an author as soon as it is finished and before it has been peer reviewed. Most news articles during this time referred to scientific data in preprints rather than conventionally peer reviewed papers. The dominant forums for these were BioRxiv (led by Richard Sever) and its medical companion MedRxiv co-founded by Harlan Krumholz.

328 *June Almeida* Details from: Meet June Almeida, the Scottish Virologist Who First Identified the Coronavirus. World from PRX. https://www.pri.org/stories/2020-05-07/meet-june-almeida-scottish-virologist-who-first-identified-coronavirus. Accessed June 9, 2020; Brocklehurst S. The Woman Who Discovered the First Coronavirus. BBC. https://www.bbc.com/news/uk-scotland-52278716. Published April 15, 2020. Accessed June 9, 2020; Gellene D. Overlooked No More: June Almeida, Scientist Who Identified the First Coronavirus. *New York Times.* https://www.nytimes.com/2020/05/08/obituaries/june-almeida-overlooked-coronavirus.html. Published May 8, 2020. Accessed June 9, 2020.

328 *took as little as thirty minutes to complete* UCSF paper: Broughton JP, Deng X, Yu G, et al. CRISPR-Cas12-based detection of SARS-CoV-2. *Nat Biotechnol.* April 2020. doi:10.1038/s41587-020-0513-4.

Broad MIT paper: Joung J, Ladha A, Saito M, et al. Point-of-care testing for COVID-19 using SHERLOCK diagnostics. *medRxiv.* May 2020. doi:10.1101/2020.05.04.20091231.

329 *Adrian Hill and Sarah Gilbert* Folegatti PM, Ewer KJ, Aley PK, et al. Safety and immunogenicity of the ChAdOx1 nCoV-19 vaccine against SARS-CoV-2: a preliminary report of a phase 1/2, single-blind, randomised controlled trial. *Lancet.* Published online July 20, 2020. doi:10.1016/S0140-6736(20)31604-4.

329 *group of engineers and epidemiologists at Yale* Peccia J, Zulli A, Brackney DE, et al. SARS-CoV-2 RNA concentrations in primary municipal sewage sludge as a leading indicator of COVID-19 outbreak dynamics. *Epidemiology.* May 2020. doi:10.1101/2020.05.19.20105999.

330 *over ninety different contaminants* Data from the CDC: Regulations: The Safe Drinking Water Act. https://www.cdc.gov/healthywater/drinking/public/regulations.html. Published October 10, 2018. Accessed June 10, 2020.

331 *Stanley Qi* Abbott TR, Dhamdhere G, Liu Y, et al. Development of CRISPR as an antiviral strategy to combat SARS-CoV-2 and influenza. *Cell.* 2020;181(4):865-876.e12.

331 *such innovation doesn't come cheap* Yates N, Métraux E, Zayner J, Johnston J, Stauffer W, Ricci DM. I have SMA. Critics of the $2 Million New Therapy Are Missing the Point. STAT. https://www.statnews.com/2019/05/31/spinal-muscular-atrophy-zolgensma-price-critics/. Published May 31, 2019. Accessed January 4, 2020; Cassidy B, Métraux E, Zayner J, Johnston J, Stauffer W, Ricci DM. How Will We Pay For Potentially Curative Gene Therapies? STAT. https://www.statnews.com/2019/06/12/paying-for-coming-generation-gene-therapies/. Published June 12, 2019. Accessed January 4, 2020.

# Index

Founded in 2017, Celadon Books, a division of Macmillan Publishers, publishes a highly curated list of twenty to twenty-five new titles a year. The list of both fiction and nonfiction is eclectic and focuses on publishing commercial and literary books and discovering and nurturing talent.